SpringerBriefs in Electrical and Computer Engineering

Control, Automation and Robotics

Series editors

Tamer Başar, Urbana, USA
Antonio Bicchi, Pisa, Italy
Miroslav Krstic, La Jolla, USA

More information about this series at http://www.springer.com/series/10059

Martin Gugat

Optimal Boundary Control and Boundary Stabilization of Hyperbolic Systems

 Birkhäuser

Martin Gugat
Mathematik
Friedrich-Alexander-Universität Erlangen-Nürnberg
Erlangen, Germany

ISSN 2191-8112   ISSN 2191-8120   (electronic)
SpringerBriefs in Electrical and Computer Engineering
ISBN 978-3-319-18889-8   ISBN 978-3-319-18890-4   (eBook)
DOI 10.1007/978-3-319-18890-4

Library of Congress Control Number: 2015939419

Mathematics Subject Classification (2010): 93C20, 93D15, 49J20

Springer Cham Heidelberg New York Dordrecht London

Printed on acid-free paper

Springer International Publishing AG Switzerland is part of Springer Science+Business Media (www.
springer.com)

Preface

For many systems in engineering, control action does not take place everywhere in the system but only at certain points, for example at the boundary. In this SpringerBrief we consider systems of this type, where the system dynamics are governed by hyperbolic partial differential equations. As a typical example we consider the wave equation, and in the chapter about nonlinear systems also the Korteweg-de Vries and Burgers equations.

The aim is to familiarize the reader with the problems of optimal boundary control and stabilization that appear in this framework. We also introduce the notion of exact controllability, which is an essential concept in this context. To keep the presentation as accessible as possible, we consider the case of systems with a state that is defined on a space interval. In the case of a finite interval that is also relevant for many applications, there are two boundary points where the system can be controlled. In the optimal control problems, the aim is to find an admissible control that is optimal. Admissible means that the control has given desired properties, for example that it steers the system to a given terminal state. Optimal means that a given objective function that reflects the preferences of the decision maker is minimal, for example the control cost given by the norm of the control on the time interval. In the stabilization problems, the aim is to define a feedback law such that for the corresponding closed loop system the state approaches a given desired trajectory, for example a position of rest exponentially fast.

This text has grown from a lecture at the Friedrich-Alexander-Universität Erlangen-Nürnberg (FAU) in the summer term 2014. The aim of the text is to give an introduction to the subject that is accessible to graduate and Ph.D. students in mathematics and engineering and research workers in the field. In order to be able to proceed quickly to optimization and stabilization results, we do not introduce the general framework of strongly continuous semigroups. Introductions to the whole field of control of infinite dimensional systems based upon the concept of semigroups are given in [4, 41, 54].

This work was supported by DFG grant CRC/Transregio 154, *Mathematical Modelling, Simulation and Optimization Using the Example of Gas Networks*, Project C03.

Erlangen, Germany   Martin Gugat

Contents

Chapter 1
Introduction

Many important systems in engineering are governed by hyperbolic partial differential equations. Typical examples are the flow through water transportation networks (see for example [31]), gas flow through pipeline networks (see for example [16] and the references therein), and power grids. Also traffic flow can be modeled by hyperbolic partial differential equations (see for example [32, 57]). These models allow to study how control action influences the states in these systems. This is the motivation to consider the following problems:

- The *optimal control problem* consists in finding a control from a given admissible set for which a given objective functional attains a value that is as small as possible. The objective function is chosen to model the preferences of the decision maker. Then the corresponding optimization problem allows a rational decision for an optimal control. The optimal control takes into account the system dynamics and depends in general on the initial state. We study optimal control problems where the admissible controls steer the system exactly to a given terminal state at a given finite time. If such controls exist, the system is called *exactly controllable*. The exact controllability properties are an important feature of hyperbolic systems.
- The aim of *stabilization* is to eliminate the effect of perturbations of the system state in order to steer the system state to a given desired trajectory. For this purpose feedback laws are introduced, that allow to react to deviations of the system state from the desired trajectory. Since the deviations are a priori unknown, the feedback laws must be well defined for all possible system states.

For a mathematical study of these problems we need an analytic framework, that ensures that solutions of the corresponding initial boundary value problems exist. On this base, we study the problems of optimal control and stabilization.

© The Author(s) 2015
M. Gugat, *Optimal Boundary Control and Boundary Stabilization of Hyperbolic Systems*, SpringerBriefs in Electrical and Computer Engineering,
DOI 10.1007/978-3-319-18890-4_1

Chapter 2
Systems governed by the wave equation

We consider systems that are governed by hyperbolic *partial differential equations* (pdes). As a first example, we consider the *wave equation*

$$y_{tt} = c^2 y_{xx}. \tag{2.1}$$

Here c is a real number and $|c|$ is called the *wave speed*. We will focus on the one-dimensional case, where we can present essential concepts. To analyze the wave equation, the concept of traveling waves is useful.

Definition 2.1. A solution of the form

$$y(t, x) = f(x - a t)$$

is called a *traveling wave* that travels with the velocity a. Here a is a real number that can be positive, negative, or zero. f denotes a real function that is defined on an interval $I, f : I \to (-\infty, \infty)$. The interval I and the velocity a determine the domain of y.

The traveling waves are solutions of a first order pde, the so-called transport equation

$$y_t = -af'(x - at) = -ay_x,$$

that is

$$y_t + ay_x = 0.$$

This can also be written as

$$(\partial_t + a\,\partial_x)\,y = 0.$$

© The Author(s) 2015   3
M. Gugat, *Optimal Boundary Control and Boundary Stabilization of Hyperbolic Systems*, SpringerBriefs in Electrical and Computer Engineering,
DOI 10.1007/978-3-319-18890-4_2

The wave equation (2.1) allows solutions that consist of waves that travel to the left-hand side and waves that travel to the right-hand side, both with the velocity $|c|$. If the corresponding operators

$$(\partial_t + c\,\partial_x)$$

and

$$(\partial_t - c\,\partial_x)$$

are multiplied, this yields

$$(\partial_t + c\,\partial_x)(\partial_t - c\,\partial_x) = (\partial_{tt} - c^2\partial_{xx})$$

that is the operator that appears in the wave equation. Here the order of the operators can be exchanged, because we also have

$$(\partial_t - c\,\partial_x)(\partial_t + c\,\partial_x) = (\partial_{tt} - c^2\partial_{xx}).$$

Thus for two given functions α and β the traveling waves

$$\alpha(x + c\,t) \quad \text{and} \quad \beta(x - c\,t)$$

solve the wave equation, provided that the derivatives α' and β' exist (for example in the sense of distributions as explained below in Chapter 7). Since the operators are linear, this implies that

$$y(t, x) = \alpha(x + c\,t) + \beta(x - c\,t)$$

solves the wave equation.

For the pde analysis often derivatives in the sense of distributions are needed, since classical solutions do not always exist. Therefore in Chapter 7 we present a very short introduction to the theory of distributions. We recommend that readers who are not familiar with the concept of the distributional derivative study Chapter 7 before they continue.

In the next sections we consider initial boundary value problems with the wave equation. We provide results about the existence and uniqueness of solutions of these initial boundary value problems. In 2.1 we consider systems with Dirichlet boundary control, in 2.2 we consider systems with Neumann boundary control, and in 2.3 we study Robin boundary control.

2.1 Dirichlet boundary control

Using the derivatives in the sense of distributions (see Chapter 7), we construct the solutions of initial boundary value problems. First we consider an initial boundary value problem with Dirichlet boundary control.

Let the length $L > 0$ of the space interval and the wave speed $c > 0$ be given. We use the standard notation

$$L^2(0, L) = \{f : [0, L] \to \mathbb{R}, f \text{ is measurable and } \int_0^L |f(x)|^2 \, dx < \infty\}.$$

Let $y_0 \in L^2(0, L)$ be given. The inequality $|f| \leq \frac{1}{2}(|f|^2 + 1)$ shows that the elements of $L^2(0, L)$ are also elements of

$$L^1(0, L) = \{f : [0, L] \to \mathbb{R}, f \text{ is measurable and } \int_0^L |f(x)| \, dx < \infty\}.$$

We use the notation

$$H^{-1}(0, L) = \{Y \in \mathcal{D}'((0, L)), \text{ there is } f \in L^2(0, L) \text{ such that } f' = Y\}.$$

The definition of the set of distributions $\mathcal{D}'((0, L))$ is given in Definition 7.2 in Chapter 7. The space $H^{-1}(0, L)$ is a *Sobolev space*. Introductions to Sobolev spaces can be found for example in [1] and [42].

Let $y_1 \in H^{-1}(0, L)$ be given. An element f of $L^2(0, L)$ with $f' = Y$ is only determined up to a constant $C \in \mathbb{R}$, since

$$(f + C)' = f'.$$

For such an antiderivative of Y (that is only determined up to an additive constant) we use the notation

$$\int_0^x Y(\sigma) \, d\sigma.$$

Let $T \in (0, \infty)$ denote a control time. Let a control $u \in L^2(0, T)$ be given. We consider the initial boundary value problem

$$(DARWP) \begin{cases} y(0, x) = y_0(x), & x \in (0, L) \\ y_t(0, x) = y_1(x), & x \in (0, L) \\ y_{tt}(t, x) = c^2 \, y_{xx}(t, x), & (t, x) \in (0, T) \times (0, L) \\ y(t, 0) = 0, & t \in (0, T) \\ y(t, L) = u(t), & t \in (0, T). \end{cases}$$

In Theorem 2.1 we present a traveling waves representation of the solution of (*DARWP*). For the infinite string, this solution is due to D'ALEMBERT.

Theorem 2.1 (see [29]). *Let $y_0 \in L^2(0, L)$, $y_1 \in H^{-1}(0, L)$ $u \in L^2(0, T)$ be given. Define the time*

$$t_0 = \frac{L}{c}$$

and the function

$$f(x) = \int_0^x y_1(\sigma)\,d\sigma \in L^2(0,L)$$

as an antiderivative of y_1. For $s \in (0, t_0)$ we define

$$\alpha_0(s) = y_0(c\,s) + \frac{1}{c}f(c\,s),$$

$$\beta_0(s) = y_0(L - c\,s) - \frac{1}{c}f(L - c\,s).$$

For sufficiently small $k \in \{1, 2, 3, \ldots\}$ and $s \in (0, t_0)$ we define

- *if k is an odd number:*

$$\alpha_k(s) = \sum_{i=0}^{(k-1)/2} 2u(s + 2it_0) - \beta_0(s),$$

$$\beta_k(s) = \sum_{i=1}^{(k-1)/2} -2u(s + (2i-1)t_0) - \alpha_0(s);$$

- *if k is an even number:*

$$\alpha_k(s) = \sum_{i=0}^{(k-2)/2} 2u(s + (2i+1)t_0) + \alpha_0(s),$$

$$\beta_k(s) = \sum_{i=0}^{(k-2)/2} -2u(s + 2it_0) + \beta_0(s).$$

Then $\alpha_k, \beta_k \in L^2(0, t_0)$. Now we define functions α and $\beta \in L^2(0, T + t_0)$.

For this purpose, for $z \in (0, \infty)$ we define the numbers $t_+(z) \in (0, t_0)$ and $j(z) \in \{0, 1, 2, 3, \ldots\}$ by the equation

$$z = t_+(z) + j(z)\,t_0.$$

For $z \in (0, T + t_0)$ we define

$$\alpha(z) = \alpha_{j(z)}(t_+(z)), \quad \beta(z) = \beta_{j(z)}(t_+(z)).$$

Then the function

$$y(t, x) = \frac{1}{2}\left[\alpha\left(t + \frac{x}{c}\right) + \beta\left(t + \frac{L - x}{c}\right)\right]$$

solves (DARWP).

Proof. First we check whether the initial conditions are satisfied. For $x \in (0, L)$ almost everywhere we have

$$\alpha \left(\frac{x}{c} \right) = \alpha_0 \left(\frac{x}{c} \right) = y_0(x) + \frac{1}{c} f(x),$$

$$\beta \left(\frac{L - x}{c} \right) = \beta_0 \left(\frac{L - x}{c} \right) = y_0(x) - \frac{1}{c} f(x).$$

This implies $y(0, x) = y_0(x)$. The velocity satisfies the equation

$$y_t(t, x) = \frac{1}{2} \left[\alpha' \left(t + \frac{x}{c} \right) + \beta' \left(t + \frac{L - x}{c} \right) \right].$$

Hence we get

$$y_t(0, x) = \frac{1}{2} \left[\alpha' \left(\frac{x}{c} \right) + \beta' \left(\frac{L - x}{c} \right) \right].$$

For $x \in (0, L)$ in the sense of distributions we have the equations

$$\alpha' \left(\frac{x}{c} \right) = \alpha_0' \left(\frac{x}{c} \right) = c y_0'(x) + f'(x),$$

$$\beta' \left(\frac{L - x}{c} \right) = \beta_0' \left(\frac{L - x}{c} \right) = -c y_0'(x) + f'(x).$$

This implies $y_t(0, x) = f'(x) = y_1(x)$.

Now we check whether y is a solution of the wave equation. We have

$$y_{tt}(t, x) = \frac{1}{2} \left[\alpha'' \left(t + \frac{x}{c} \right) + \beta'' \left(t + \frac{L - x}{c} \right) \right],$$

$$y_x(t, x) = \frac{1}{2c} \left[\alpha' \left(t + \frac{x}{c} \right) - \beta' \left(t + \frac{L - x}{c} \right) \right],$$

$$y_{xx}(t, x) = \frac{1}{2c^2} \left[\alpha'' \left(t + \frac{x}{c} \right) + \beta'' \left(t + \frac{L - x}{c} \right) \right] = \frac{1}{c^2} y_{tt}(t, x).$$

Now we look at the boundary conditions. We start with $x = 0$. For $t \in [0, T]$ we have

$$y(t, 0) = \frac{1}{2} [\alpha(t) + \beta(t + t_0)]$$

$$= \frac{1}{2} [\alpha_{j(t)}(t - j(t)t_0) + \beta_{j(t+t_0)}(t - j(t)t_0)].$$

We have $j(t + t_0) = j(t) + 1$.

Case 1: $j(t)$ **is an even number** Then $j(t + t_0)$ is an odd number. Hence

$$
y(t, 0) = \frac{1}{2} \left[\sum_{i=0}^{(j(t)-2)/2} 2u(t - j(t)t_0 + (2i + 1)t_0) + \alpha_0(t - j(t)t_0) \right.
$$

$$
\left. + \sum_{i=1}^{(j(t+t_0)-1)/2} -2u(t - j(t)t_0 + (2i - 1)t_0) - \alpha_0(t - j(t)t_0) \right]
$$

$$
= \sum_{i=1}^{j(t)/2} u(t - j(t)t_0 + (2(i - 1) + 1)t_0)
$$

$$
+ \sum_{i=1}^{(j(t+t_0)-1)/2} -u(t - j(t)t_0 + (2i - 1)t_0)
$$

$$
= 0.
$$

Case 2: $j(t)$ **is a odd number** Then $j(t + t_0)$ is an even number. Thus

$$
y(t, 0) = \frac{1}{2} \left[\sum_{i=0}^{(j(t)-1)/2} 2u(t - j(t)t_0 + 2it_0) - \beta_0(t - j(t)t_0) \right.
$$

$$
\left. + \sum_{i=0}^{(j(t+t_0)-2)/2} -2u(t - j(t)t_0 + 2it_0) + \beta_0(t - j(t)t_0) \right]
$$

$$
= \sum_{i=0}^{(j(t)-1)/2} u(t - j(t)t_0 + 2it_0) - u(t - j(t)t_0 + 2it_0)
$$

$$
= 0.
$$

Thus we have shown $y(t, 0) = 0$ for $t \in (0, T)$ almost everywhere.

Now we consider the boundary point $x = L$. We have

$$
y(t, L) = \frac{1}{2} [\alpha(t + t_0) + \beta(t)]
$$

$$
= \frac{1}{2} [\alpha_{j(t+t_0)}(t - j(t)t_0) + \beta_{j(t)}(t - j(t)t_0)].
$$

We have $j(t + t_0) = j(t) + 1$.

Case 1: $j(t)$ **is an even number** Then $j(t + t_0)$ is an odd number. Thus we get

$$
y(t, L) = \frac{1}{2} \left[\sum_{i=0}^{(j(t)+1-1)/2} 2u(t - j(t)t_0 + 2it_0) - \beta_0(t - j(t)t_0) \right.
$$

$$
\left. + \sum_{i=0}^{(j(t)-2)/2} -2u(t - j(t)t_0 + 2i\,t_0) + \beta_0(t - j(t)t_0) \right]
$$

$$
= \sum_{i=0}^{j(t)/2} u(t - j(t)t_0 + 2it_0) + \sum_{i=0}^{(j(t)/2-1)} -u(t - j(t)t_0 + 2it_0)
$$

$$
= u(t - j(t)t_0 + j(t)t_0)
$$

$$
= u(t).
$$

Case 2: $j(t)$ **is an odd number** Then $j(t + t_0)$ is an even number. Hence we have

$$
y(t, L) = \frac{1}{2} \left[\sum_{i=0}^{(j(t)+1-2)/2} 2u(t - j(t)t_0 + (2i + 1)t_0) + \alpha_0(t - j(t)t_0) \right.
$$

$$
\left. + \sum_{i=1}^{(j(t)-1)/2} -2u(t - j(t)t_0 + (2i - 1)t_0) - \alpha_0(t - j(t)t_0) \right]
$$

$$
= \sum_{i=0}^{(j(t)-1)/2} u(t - j(t)t_0 + (2i + 1)t_0)
$$

$$
- \sum_{i=1}^{(j(t)-1)/2} u(t - j(t)t_0 + (2i - 1)t_0)
$$

$$
= \sum_{i=0}^{(j(t)-1)/2} u(t - j(t)t_0 + (2i + 1)t_0)
$$

$$
- \sum_{i=0}^{(j(t)-1)/2-1} u(t - j(t)t_0 + (2(i + 1) - 1)t_0)
$$

$$
= u(t - j(t)t_0 + (j(t) - 1 + 1)t_0)
$$

$$
= u(t).
$$

Thus we have $y(t, L) = u(t)$ for $t \in (0, T)$ almost everywhere.

So we have shown that $y(t, x)$ solves (DARWP) and Theorem 2.1 is proved.$\square$

Remark 2.1. The solution can also be represented as a Fourier series, see Remark 2.2. An advantage of the representation of $y(t, x)$ from Theorem 2.1 is that in contrast to an infinite series it can be implemented on a computer, also for initial data y_0, y_1 with kinks and jumps. The number $j(z)$ from the equation $z = t_+(z) + j(z) t_0$ is computed as the integer part of z/t_0.

Another advantage of the representation of $y(t, x)$ from Theorem 2.1 is that it illustrates very clearly the limited **domains of dependence** of the values of $y(t, x)$ of the control values u and the initial data y_0, y_1.

Similarly it allows to determine the limited **domains of influence** where the values of the solutions are influenced by the initial values $y_0(x)$ and $y_1(x)$.

Exercise 2.1. Produce an implementation of the representation of the solution $y(y_0, y_1, u)$ from Theorem 2.1!

Remark 2.2. It is also possible to control the system at both ends of the string: This corresponds to Dirichlet boundary conditions

$$y(0, t) = u_{-1}(t), \quad y(L, t) = u_1(t), \quad t \in (0, \infty). \tag{2.2}$$

Here the control action is given by two control functions u_{-1} and u_1.

The solution can also be represented as a Fourier series.

For this purpose, first the initial data must be represented as a Fourier series:

$$y_0(x) = \sum_{j=1}^{\infty} a_j \sin\left(\frac{j\pi x}{L}\right), \quad y_1(x) = \sum_{j=1}^{\infty} b_j \frac{j\pi c}{L} \sin\left(\frac{j\pi x}{L}\right).$$

This yields the solution y of the initial boundary value problem with the boundary conditions (2.2) in the form

$$y(t, x) = \sum_{j=1}^{\infty} \sin\left(\frac{j\pi x}{L}\right) \left[a_j \cos\left(\frac{j\pi ct}{L}\right) + b_j \sin\left(\frac{j\pi ct}{L}\right)\right]$$

$$+ \sum_{j=1}^{\infty} \sin\left(\frac{j\pi x}{L}\right) \left[\frac{2c}{L} \int_0^t [u_{-1}(s) - (-1)^j u_1(s)] \sin\left(\frac{c\pi j}{L}(t - s)\right) ds\right].$$

With $u_{-1}(s) = 0$ this yields again our solution from Theorem 2.1.

Exercise 2.2. Show that with $u_{-1}(s) = 0$ the above series yields the solution from Theorem 2.1.

Remark 2.3. D'Alembert gave his solution of the wave equation in *Recherches sur la courbe que forme une corde tendue mise en vibration, Mem. Acad. Sci. Berlin 3, 214–219, (1747)*. The discussion about the different representation of solutions of the wave equation as a series or a sum of traveling waves has a long history, see [17].

Example 2.1. Let $k \in \{1, 2, 3, \ldots\}$ be given. For the initial boundary value problem with the initial state

$$y_0(x) = \sin\left(\frac{k\pi x}{L}\right), \quad y_1(x) = 0 \tag{2.3}$$

and the boundary conditions $u_{-1}(t) = u_1(t) = 0$ we get the Fourier series of the solution

$$y(t, x) = \sin\left(\frac{k\pi x}{L}\right) \cos\left(\frac{k\pi ct}{L}\right). \tag{2.4}$$

This representation shows that in certain points (namely the roots of $\sin(k\pi x/L)$) the solution remains zero at all times. These points are called *nodes*.

Trigonometric identities yield the solution in the form

$$y(t, x) = \frac{1}{2}\left[\sin\left(\frac{k\pi}{L}(x + ct)\right) + \sin\left(\frac{k\pi}{L}(x - ct)\right)\right]. \tag{2.5}$$

This corresponds to the representation that is given in Theorem 2.1. Here we have

$$\alpha_0(t) = y_0(ct) = \sin\left(\frac{k\pi ct}{L}\right),$$

$$\beta_0(t) = y_0(L - ct) = \sin\left(k\pi - \frac{k\pi ct}{L}\right) = (-1)^{k+1}\sin\left(\frac{k\pi ct}{L}\right) = -(-1)^k \alpha_0(t).$$

Exercise 2.3. Let $y_0(x) = x$, $y_1(x) = 0$.
For $t \in (0, 2t_0)$ and $l \in \{0, 1, \ldots, K - 1\}$ let

$$u(t + 2l t_0) = \frac{1}{2K}(L - ct)$$

and for $t > 2K t_0$ let $u(t) = 0$.

Determine for the corresponding solution $y(t, x)$ von $(DARWP)$ the values

$$(y(Kt_0, x), y_t(Kt_0, x)) \quad \text{and} \quad (y(2K t_0, x), y_t(2K t_0, x)).$$

Solution of Exercise 2.3. *We use Theorem 2.1. We have $f(x) = 0$ and*

$$\alpha_0(s) = y_0(cs) = cs \quad \text{and} \quad \beta_0(s) = y_0(L - cs) = L - cs.$$

If k is odd, for all $s \in (0, t_0)$ we have

$$\alpha_k(s) = \sum_{i=0}^{(k-1)/2} 2u(s + 2i t_0) - \beta_0(s)$$

$$= \sum_{i=0}^{(k-1)/2} \frac{1}{K}(L - cs) - (L - cs)$$

$$= (k+1)/2 \, \frac{1}{K} \, (L - cs) - (L - cs)$$

$$= (L - cs) \left[\frac{k+1-2K}{2K} \right]$$

$$\beta_k(s) = \sum_{i=1}^{(k-1)/2} -2u(s + (2i-1)t_0) - \alpha_0(s)$$

$$= \sum_{i=1}^{(k-1)/2} -2u(s + t_0 + (2i-2)t_0) - \alpha_0(s)$$

$$= \sum_{i=1}^{(k-1)/2} -\frac{1}{K}(L - c(s + t_0)) - cs$$

$$= (k-1)/2 \frac{1}{K} cs - cs$$

$$= \left[\frac{k-1-2K}{2K} \right] cs.$$

If k is even, we have

$$\alpha_k(s) = \sum_{i=0}^{(k-2)/2} 2u(s + (2i+1)t_0) + \alpha_0(s)$$

$$= - \sum_{i=0}^{(k-2)/2} \frac{1}{K}(cs) + c\,s$$

$$= -\frac{k}{2}\frac{1}{K}(cs) + cs$$

$$= \frac{2K - k}{2K}(cs),$$

$$\beta_k(s) = \sum_{i=0}^{(k-2)/2} -2u(s + 2it_0) + \beta_0(s)$$

$$= \sum_{i=0}^{(k-2)/2} -\frac{1}{K}(L - cs) + (L - cs)$$

$$= -\frac{k}{2}\frac{1}{K}(L - cs) + (L - cs)$$

$$= \frac{2K - k}{2K}(L - cs).$$

Then α_k and β_k are in $L^2(0, t_0)$. Theorem 2.1 implies

$$y(K t_0, x) = \frac{1}{2}\left[\alpha\left(K t_0 + \frac{x}{c}\right) + \beta\left(K t_0 + \frac{L-x}{c}\right)\right]$$

$$= \frac{1}{2}\left[\alpha_K\left(\frac{x}{c}\right) + \beta_K\left(\frac{L-x}{c}\right)\right]$$

If K is even, this yields

$$y(K t_0, x) = \frac{1}{2}\left[\frac{2K - K}{2K}x + \frac{2K - K}{2K}x\right] = \frac{1}{2}x$$

and

$$y_t(K t_0, x) = \frac{1}{2}\left[\frac{2K - K}{2K}c - \frac{2K - K}{2K}c\right] = 0.$$

If K is an odd number, we get

$$y(K t_0, x) = \frac{1}{2}\left[(L - x)\left[\frac{K + 1 - 2K}{2K}\right] + \left[\frac{K - 1 - 2K}{2K}\right](L - x)\right]$$

$$= \frac{1}{2}(L - x)[-1] = \frac{1}{2}(x - L)$$

and

$$y_t(K t_0, x) = \frac{1}{2}\left[-c\left[\frac{K + 1 - 2K}{2K}\right] + c\left[\frac{K - 1 - 2K}{2K}\right]\right]$$

$$= \frac{1}{2}\left[-c\frac{2}{2K}\right] = -\frac{c}{K}.$$

Moreover, we have

$$y(2 K t_0, x) = \frac{1}{2}\left[\frac{2K - 2K}{2K}cx + \frac{2K - 2K}{2K}cx\right] = 0$$

and

$$y_t(K t_0, x) = \frac{1}{2}\left[\frac{2K - 2K}{2K}c - \frac{2K - 2K}{2K}c\right] = 0.$$

Thus we have solved Exercise 2.3. The control u steers the system to a position of rest at the time $2Kt_0$. Afterwards with zero control the system remains in this position. At the time $K t_0$ for even K the velocity is zero and the position is half the initial position.

Now we look at the uniqueness of the solution.

Theorem 2.2 (Uniqueness). *Let $y_0 \in L^2(0, L)$, $y_1 \in H^{-1}(0, L)$, and $u \in L^2(0, T)$ be given. Then the solution of the initial boundary value problem (DARWP) is uniquely determined.*

Proof. Let z_1 and z_2 denote solutions of (DARWP). We consider the difference

$$y = z_2 - z_1.$$

Due to the linearity of the system y solves the homogeneous initial boundary value problem

$$\begin{cases} y(0, x) = 0, & x \in (0, L) \\ y_t(0, x) = 0, & x \in (0, L) \\ y_{tt}(t, x) = c^2 \, y_{xx}(t, x), & (t, x) \in (0, T) \times (0, L) \\ y(t, 0) = 0, & t \in (0, T) \\ y(t, L) = 0, & t \in (0, T). \end{cases}$$

We consider the energy

$$E(t) = \frac{1}{2} \int_0^L y(t, x)^2 + \frac{1}{c^2} \int_0^L \left(\int_0^x y_t(t, s) \, ds - \frac{1}{L} \int_0^L \int_0^z y_t(t, s) \, ds \, dz \right)^2 dx \tag{2.6}$$

(see [22]). Then the value of $E(t)$ is independent of the choice of the antiderivative of y_t. We define

$$Y(t, x) = \int_0^x y_t(t, s) \, ds - \frac{1}{L} \int_0^L \int_0^z y_t(t, s) \, ds \, dz.$$

Then the definition of Y implies $\int_0^L Y(t, x) \, dx = 0$.
 For $t > 0$ the definition of Y and the wave equation imply the equation

$$Y(t, x) - Y(0, x)$$

$$= \int_0^x [y_t(t, s) - y_t(0, s)] \, ds - \frac{1}{L} \int_0^L \int_0^z [y_t(t, s) - y_t(0, s)] \, ds \, dz$$

$$= \int_0^x \int_0^t y_{tt}(\tau, s) \, d\tau \, ds - \frac{1}{L} \int_0^L \int_0^z \int_0^t y_{tt}(\tau, s) \, d\tau \, ds \, dz$$

$$= c^2 \int_0^t \int_0^x y_{xx}(\tau, s) \, ds \, d\tau - \frac{1}{L} c^2 \int_0^L \int_0^t \int_0^z y_{xx}(\tau, s) \, ds \, d\tau \, dz.$$

Now we can evaluate the integrals and with the homogeneous Dirichlet boundary conditions we get the equation

$$Y(t, x) - Y(0, x)$$

$$= c^2 \int_0^t [y_x(\tau, x) - y_x(\tau, 0)] \, d\tau - c^2 \frac{1}{L} \int_0^L \int_0^t [y_x(\tau, z) - y_x(\tau, 0)] \, d\tau \, dz$$

$$= c^2 \int_0^t y_x(\tau, x) \, d\tau - c^2 \frac{1}{L} \int_0^t \int_0^L [y_x(\tau, z) - y_x(\tau, 0)] \, dz \, d\tau - c^2 \int_0^t y_x(\tau, 0) \, d\tau$$

$$= c^2 \int_0^t y_x(\tau, x) \, d\tau - c^2 \frac{1}{L} \int_0^t [y(\tau, L) - y(\tau, 0)] \, d\tau$$

$$+ c^2 \int_0^t y_x(\tau, 0) \, d\tau - c^2 \int_0^t y_x(\tau, 0) \, d\tau$$

$$= c^2 \int_0^t y_x(\tau, x) \, d\tau.$$

This yields the equation

$$Y_t = c^2 \, y_x$$

in the sense of distributions. With the definition of Y, integration by parts and the equation $Y_x = y_t$ this yields

$$0 = \int_0^L y(t, x) \, y_t(t, x) - y_t(t, x) \, y(t, x) \, dx$$

$$= \int_0^L y(t, x) \, y_t(x, t) - Y_x(t, x) \, y(t, x) \, dx$$

$$= \int_0^L y(t, x) \, y_t(t, x) + Y(t, x) \, y_x(t, x) \, dx$$

$$= \int_0^L y(t, x) \, y_t(t, x) + \frac{1}{c^2} Y(t, x) Y_t(t, x) \, dx.$$

This implies

$$E(t) - E(0) = \frac{1}{2} \int_0^L \left[y^2(\tau, x) + \frac{1}{c^2} Y^2(\tau, x) \right] |_{\tau=0}^t \, dx$$

$$= \int_0^L \int_0^t y(\tau, x) \, y_t(\tau, x) + \frac{1}{c^2} Y(\tau, x) \, Y_t(\tau, x) \, d\tau \, dx$$

$$= \int_0^t \int_0^L y(\tau, x) \, y_t(\tau, x) + \frac{1}{c^2} Y(\tau, x) \, Y_t(\tau, x) \, dx \, d\tau$$

$$= 0.$$

Thus for all $t > 0$ we get

$$E(t) = E(0) = 0.$$

Hence $y(t, x) = 0$ and Theorem 2.2 is proven. $\square$

Remark 2.4. The proof of Theorem 2.2 shows that also for arbitrary initial states $y_0 \in L^2(0, L)$, $y_1 \in H^{-1}(0, L)$ with homogeneous Dirichlet boundary conditions (that is with $u = 0$) the energy is conserved, that is for all $t > 0$ we have

$$E(t) = E(0).$$

Remark 2.5. In the two-dimensional case the wave equation has the form

$$v_{tt} = c^2 \left(v_{xx} + v_{yy} \right).$$

This is for example a model for a vibrating membrane as in a loudspeaker.

Also in the 2-d case the solution of the initial boundary value problem can be represented as a series where the terms corresponds to the eigenfunctions of the Laplacian that are the solutions of the sequence of equations

$$(\varphi_j)_{xx} + (\varphi_j)_{yy} + \lambda_j \varphi_j = 0.$$

Here the λ_j are the corresponding eigenvalues.

A solution in the whole plane can also be represented in the integral form

$$v(t, x, y) = \int_0^\pi \alpha(ct + \cos(\varphi)\,x + \sin(\varphi)\,y, \ \varphi) + \beta(ct - \cos(\varphi)\,x - \sin(\varphi)\,y, \ \varphi)\,d\mu(\varphi).$$

where μ is a Lebesgue-Stieltjes measure.

In dimension three the wave equation has the form

$$v_{tt} = c^2 \left(v_{xx} + v_{yy} + v_{zz} \right).$$

This equation appears for example as a model for sound waves. Here in general the solution of the initial boundary value problem has to be computed numerically.

2.2   Neumann boundary control

Now we consider problems with Neumann-boundary conditions. To study this problem, we need the Sobolev space $H^1(0, L)$ that we define as

$$H^1(0, L) = \{ f \in L^2(0, L) :$$

For the derivative in the sense of distributions we have $f' \in L^2(0, L)$}.

Let $y_0 \in H^1(0, L)$ and $y_1 \in L^2(0, L)$ be given. Let $T \in (0, \infty)$ denote a given terminal time. For a given control $u \in L^2(0, T)$ we consider the initial boundary value problem

$$(NARWP) \begin{cases} y(0, x) = y_0(x), & x \in (0, L) \\ y_t(0, x) = y_1(x), & x \in (0, L) \\ y_{tt}(t, x) = c^2\, y_{xx}(t, x), & (t, x) \in (0, T) \times (0, L) \\ y(t, 0) = 0, & t \in (0, T) \\ y_x(t, L) = u(t), & t \in (0, T). \end{cases}$$

Theorem 2.3. *Assume that* $y_0 \in H^1(0, L)$ *with* $y_0(0) = 0$, $y_1 \in L^2(0, L)$ *and* $u \in L^2(0, T)$. *Let* $t_0 = \frac{L}{c}$. *For* $n \in \{0, 1, 2, \ldots\}$ *define*

$$\varphi_n(x) = \frac{\sqrt{2}}{\sqrt{L}} \sin\left(\left(\frac{\pi}{2} + n\pi\right)\frac{x}{L}\right),$$

$$\alpha_n^0 = \int_0^L y_0(x)\, \varphi_n(x)\, dx,$$

$$\alpha_n^1 = \int_0^L y_1(x)\, \varphi_n(x)\, dx,$$

$$\alpha_n(t) = \alpha_n^0 \cos\left(\left(\frac{\pi}{2} + n\pi\right)\frac{t}{t_0}\right) + \alpha_n^1 \frac{1}{\frac{1}{t_0}\left(\frac{\pi}{2} + n\pi\right)} \sin\left(\left(\frac{\pi}{2} + n\pi\right)\frac{t}{t_0}\right)$$

$$+ (-1)^n c^2\, \frac{\sqrt{2}}{\sqrt{L}} \left[\frac{1}{t_0}\left(\frac{\pi}{2} + n\pi\right)\right]^{-1} \int_0^t u(s) \sin\left(\left(\frac{\pi}{2} + n\pi\right)\frac{(t-s)}{t_0}\right) ds.$$

Then

$$y(t, x) = \sum_{n=0}^{\infty} \alpha_n(t)\, \varphi_n(x)$$

is a solution of (NARWP). *For all* $t \in (0, T)$ *we have*

$$y(t, \cdot) \in L^2(0, L)$$

and

$$\int_0^L y(t, x)^2\, dx = \sum_{n=0}^{\infty} (\alpha_n(t))^2.$$

Moreover

$$\left(\sum_{n=0}^{\infty}(\alpha_n(t))^2\right)^{1/2} \le \left(\sum_{n=0}^{\infty}(\alpha_n^0)^2\right)^{1/2} + \left(\sum_{n=0}^{\infty}\left(\frac{\alpha_n^1}{\frac{1}{t_0}\left(\frac{\pi}{2}+n\pi\right)}\right)^2\right)^{1/2}$$

$$+ c^2\frac{\sqrt{2}}{\sqrt{L}}\left[\sum_{n=0}^{\infty}\left(\frac{1}{\frac{1}{t_0}\left(\frac{\pi}{2}+n\pi\right)}\right)^2\left(\int_0^t u(s)\sin\left(\frac{1}{t_0}\left(\frac{\pi}{2}+n\pi\right)(t-s)\right)ds\right)^2\right]^{1/2}.$$

In particular we have

$$y \in L^{\infty}((0,T), L^2(0,L))$$

and even

$$y \in C((0,T), L^2(0,L)).$$

For the partial derivatives of y we have

$$y_t(t,x) = \sum_{n=0}^{\infty}\alpha_n'(t)\,\varphi_n(x), \quad y_x(t,x) = \sum_{n=0}^{\infty}\alpha_n(t)\,\varphi_n'(x).$$

This implies

$$y \in L^{\infty}((0,T), H^1(0,L)).$$

In fact, we even have

$$y \in C((0,T), H^1(0,L)) \cap C^1((0,T), L^2(0,L)).$$

Proof. In order to obtain a series representation of the solution, we consider the eigenvalue problem

$$\varphi_{xx}(x) = -\lambda\,\varphi(x), \quad x \in [0,L], \quad \varphi(0) = 0, \quad \varphi_x(L) = 0.$$

The corresponding eigenfunctions are

$$\varphi_n(x) = \frac{\sqrt{2}}{\sqrt{L}}\sin((\frac{\pi}{2} + n\pi)\frac{x}{L}), \quad n \in \{0,1,2,\ldots\} \qquad (2.7)$$

with the eigenvalues

$$\lambda_n = \frac{1}{L^2}\left(\frac{\pi}{2} + n\pi\right)^2$$

and the normalization

$$\int_0^L \varphi_n(x)^2\,dx = 1, \quad n \in \{0,1,2,\ldots\}.$$

The sequence of functions $((\varphi_n(x))_{n=0}^{\infty}$ is a complete orthonormal system in the Hilbert space $L^2(0, L)$ (see spectral theory of symmetric compact operators, Sturm–Liouville eigenvalue problems, for example [5]).

Now we want to determine a sequence of functions $\alpha_n(t) : [0, T] \to \mathbb{R}$ such that the solution of (NARWP) can be represented by the following series:

$$y(t, x) = \sum_{n=0}^{\infty} \alpha_n(t)\, \varphi_n(x). \tag{2.8}$$

For all $n \in \{0, 1, 2, \ldots\}$ we have

$$\int_0^L y_{tt}(t, x)\, \varphi_n(x)\, dx$$

$$= c^2 \int_0^L y_{xx}(t, x)\, \varphi_n(x)\, dx$$

$$= -c^2 \int_0^L y_x(t, x)\, (\varphi_n)_x(x)\, dx + c^2\, y_x(t, x)\, \varphi_n(x)|_{x=0}^L$$

$$= c^2 \int_0^L y(t, x)\, (\varphi_n)_{xx}(x)\, dx - c^2\, y(t, x)\, (\varphi_n)_x(x)|_{x=0}^L$$

$$+ c^2\, y_x(t, x)\, \varphi_n(x)|_{x=0}^L$$

$$= c^2 \int_0^L y(t, x)\, (\varphi_n)_{xx}(x)\, dx + c^2\, u(t)\, \varphi_n(L)$$

$$= -c^2\, \lambda_n \int_0^L y(t, x)\, \varphi_n(x)\, dx + u(t)\, (-1)^n c^2\, \frac{\sqrt{2}}{\sqrt{L}}.$$

This yields the sequence of differential equations

$$\alpha_n''(t) = -\frac{c^2}{L^2}\left(\frac{\pi}{2} + n\pi\right)^2 \alpha_n(t) + (-1)^n\, c^2\, \frac{\sqrt{2}}{\sqrt{L}}\, u(t)$$

with the initial conditions

$$\alpha_n(0) = \int_0^L y_0(x)\, \varphi_n(x)\, dx =: \alpha_n^0, \quad \alpha_n'(0) = \int_0^L y_1(x)\, \varphi_n(x)\, dx =: \alpha_n^1.$$

The solutions are

$$\alpha_n(t) = \alpha_n^0 \cos\left(\frac{c}{L}\left(\frac{\pi}{2} + n\pi\right) t\right) + \alpha_n^1 \frac{1}{\frac{c}{L}\left(\frac{\pi}{2} + n\pi\right)} \sin\left(\frac{c}{L}\left(\frac{\pi}{2} + n\pi\right) t\right)$$

$$+ (-1)^n c^2\, \frac{\sqrt{2}}{\sqrt{L}}\left[\frac{c}{L}\left(\frac{\pi}{2} + n\pi\right)\right]^{-1} \int_0^t u(s) \sin\left(\frac{c}{L}\left(\frac{\pi}{2} + n\pi\right)(t - s)\right) ds.$$

With these functions $\alpha_n(t)$ the solution $y(t, x)$ of (NARWP) has the series representation (2.8). Thus Theorem 2.3 is proved. $\square$

Remark 2.6. Theorem 2.3 is similar to Theorem 2.53 (p. 68) in Jean-Michel Coron's book [10]. We have

$$\varphi_n'(x) = \frac{\sqrt{2}}{L\sqrt{L}} \left(\frac{\pi}{2} + n\pi\right) \cos\left(\left(\frac{\pi}{2} + n\pi\right)\frac{x}{L}\right).$$

With the boundary conditions this implies for $k \neq n$

$$\int_0^L \varphi_k'(x)\, \varphi_n'(x)\, dx = 0.$$

Therefore in this case (i.e., with these boundary conditions) also the derivatives $(\varphi_n')_n$ form an orthogonal system in $L^2(0, L)$.

Now we come to the question of the uniqueness of the solution. We first look at initial states $y_1 \in H^1(0, L)$ and $y_0 \in H^2(0, L)$, where $H^2(0, L)$ that is defined as

$$H^2(0, L) = \{f \in L^2(0, L) :$$

For the derivative in the sense of distributions we have $f'' \in L^2(0, L)\}$.

Theorem 2.4 (Uniqueness for regular initial data). *Let $y_0 \in H^2(0, L)$ with $y_0(0) = 0$, $y_1 \in H^1(0, L)$ with $y_1(0) = 0$, $u \in H^1(0, T)$ with $u(0) = \partial_x(y_0(L))$ and $u'(0) = y_1'(L)$ be given. Then the solution of the initial boundary value problem (NARWP) in $H^2((0, T) \times (0, L))$ is uniquely determined.*

In particular this implies that for $y_0 = y_1 = 0$ and $u = 0$ the unique solution of (NARWP) is $y(t, x) = 0$.

Proof. Let z_1 and z_2 denote solutions of (NARWP). We consider the difference

$$y = z_2 - z_1.$$

Then y solves the homogeneous initial boundary value problem

$$\begin{cases} y(0, x) = 0, & x \in (0, L) \\ y_t(0, x) = 0, & x \in (0, L) \\ y_{tt}(t, x) = c^2\, y_{xx}(t, x), & (t, x) \in (0, T) \times (0, L) \\ y(t, 0) = 0, & t \in (0, T) \\ y_x(t, L) = 0, & t \in (0, T). \end{cases}$$

For $t > 0$ we have $y_x^2(t, \cdot)$, $y_t^2(t, \cdot) \in H^1(0, L)$.

We consider the classical energy

$$E(t) = \frac{1}{2} \int_0^L y_x(t, x)^2 + \frac{1}{c^2}\, (y_t(t, x))^2\, dx.$$

We have

$$
\begin{aligned}
E(t) - E(0) &= \frac{1}{2} \int_0^L \left[y_x^2(\tau, x) + \frac{1}{c^2} y_t(\tau, x)^2 \right] \Big|_{\tau=0}^t dx \\
&= \int_0^L \int_0^t y_x(\tau, x) y_{xt}(\tau, x) + \frac{1}{c^2} y_t(\tau, x) y_{tt}(\tau, x) \, d\tau \, dx \\
&= \int_0^L \int_0^t y_x(\tau, x) y_{xt}(\tau, x) + y_t(\tau, x) y_{xx}(\tau, x) \, d\tau \, dx \\
&= \int_0^L \int_0^t y_x(\tau, x) y_{xt}(\tau, x) - y_{tx}(\tau, x) y_x(\tau, x) \, dx \, d\tau \\
&\quad + \int_0^t y_t(\tau, x) y_x(\tau, x)|_{x=0}^L \, d\tau \\
&= 0.
\end{aligned}
$$

This implies the equation

$$
E(t) = E(0) = 0.
$$

Hence we have $y(t, x) = 0$ and Theorem 2.4 is proved. $\square$

Remark 2.7. In the proof of Theorem 2.4 we have shown that the equations $u(t) = 0 = \partial_x (y_0(L))$ implies that the energy $E(t)$ remains constant, that is it is conserved.

Exercise 2.4. Let $f \in H^1(0, 1)$ be given. Show that $\partial_x(f^2(x)) = 2f(x)f'(x)$.

Solution of Exercise 2.4.

Step 1: For all test functions $\varphi \in \mathcal{D}((0, 1))$ Lemma 7.2 (Product rule) implies

$$
\partial_x(f\varphi) = f'\varphi + f\varphi'.
$$

In our case we have

$$
\partial_x(f\varphi) = f'\varphi + f\varphi' \in L^2(0, 1).
$$

Step 2: For all test functions $\varphi \in \mathcal{D}((0, 1))$ we have

$$
\begin{aligned}
&\langle \partial_x(f^2(x)), \varphi(x) \rangle - \langle f(x)f'(x), \varphi(x) \rangle \\
&= -\langle f(x)f'(x), \varphi(x) \rangle - \langle f^2(x), \varphi'(x) \rangle \\
&= -\int_0^1 f(x)f'(x)\varphi(x) \, dx - \int_0^1 f^2(x)\varphi'(x) \, dx \\
&= -\int_0^1 f(x)\left[f(x)'\varphi(x) + f(x)\varphi'(x) \right] dx
\end{aligned}
$$

$$= - \int_0^1 f(x) \, \partial_x \left[f(x) \, \varphi(x) \right] dx$$

$$= \int_0^1 f'(x) \left[f(x) \, \varphi(x) \right] dx$$

$$= \langle f(x) f'(x), \, \varphi(x) \rangle.$$

This yields the assertion. Thus we have solved Exercise 2.4.

Now we can also show the uniqueness of the solution of (*NARWP*) with weaker regularity assumptions than in Theorem 2.4.

Theorem 2.5 (Uniqueness in the general case). *Let $y_0 \in H^1(0, L)$ with $y_0(0) = 0$, $y_1 \in L^2(0, L)$ and $u \in L^2(0, T)$ be given. Then the solution of the initial boundary value problem (NARWP) in $L^\infty((0, T), H^1((0, L))$ is uniquely determined.*

Proof. Let z_1 and z_2 denote solutions of (*NARWP*). We consider the difference

$$y = z_2 - z_1.$$

Then y solves the homogeneous initial boundary value problem

$$\begin{cases} y(0, x) = 0, & x \in (0, L) \\ y_t(0, x) = 0, & x \in (0, L) \\ y_{tt}(t, x) = c^2 \, y_{xx}(t, x), & (t, x) \in (0, T) \times (0, L) \\ y(t, 0) = 0, & t \in (0, T) \\ y_x(t, L) = 0, & t \in (0, T). \end{cases}$$

Theorem 2.4 implies $y(t, x) = 0$. Thus we have shown Theorem 2.5. □

Exercise 2.5. Show that the assumptions of Theorem 2.3 imply

$$y \in C((0, T), H^1(0, L)) \cap C^1((0, T), L^2(0, L)).$$

2.3   Robin boundary control

In this section we consider the wave equation with Robin boundary control at $x = L$. Let $y_0 \in H^1(0, L)$ with $y_0(0) = 0$, $y_1 \in L^2(0, L)$ and $u \in L^2(0, T)$ be given. Let $\gamma \in (-1/L, 0)$. We consider the initial boundary value problem

$$(RARWP) \begin{cases} y(0, x) & = y_0(x), & x \in (0, L) \\ y_t(0, x) & = y_1(x), & x \in (0, L) \\ y_{tt}(t, x) & = c^2 \, y_{xx}(t, x), & (t, x) \in (0, T) \times (0, L) \\ y(t, 0) & = 0, & t \in (0, T) \\ y_x(t, L) - \gamma \, y(t, L) & = u(t), & t \in (0, T). \end{cases}$$

The boundary condition at $x = L$ is named ROBIN-boundary condition after VICTOR GUSTAVE ROBIN (1855–1897). A typical application of Robin boundary conditions is the heat transfer at the boundary of a body

$$y_x(t, L) = \gamma [y(t, L) - y^*(t, L)],$$

where $y^*(t, L)$ denotes the temperature around the body. The case $\gamma = 0$ corresponds to an isolated boundary. For $\gamma \to \infty$ the Robin boundary condition approaches the Dirichlet case

$$y(t, L) = y^*(t, L),$$

where the temperature at the boundary is equal to the temperature around the body.

Again we consider the question of existence and uniqueness of solutions for the initial boundary value problem. As in Theorem 2.3 we can look at a series representation of the solution.

Theorem 2.6. *Let $y_0 \in H^1(0, L)$ with $y_0(0) = 0$, $y_1 \in L^2(0, L)$, $\gamma \in (-1/L, 0)$, and $u \in H^1(0, T)$ be given. Define $t_0 = \frac{L}{c}$.*

For $n \in \{0, 1, 2, \ldots\}$ let $\varphi_n(x)$ denote the solutions of the eigenvalue problem

$$\varphi_{xx}(x) = -\lambda \, \varphi(x), \ x \in [0, L], \ \varphi(0) = 0, \ \varphi_x(L) = \gamma \, \varphi(L)$$

with the eigenvalues

$$\lambda_n = \frac{\omega_n^2}{L^2},$$

$$\alpha_n^0 = \int_0^L y_0(x) \, \varphi_n(x) \, dx,$$

$$\alpha_n^1 = \int_0^L y_1(x) \, \varphi_n(x) \, dx,$$

$$\alpha_n(t) = \alpha_n^0 \cos \left(\frac{c}{L} \omega_n t \right) + \alpha_n^1 \frac{1}{\frac{c}{L} \omega_n} \sin \left(\frac{c}{L} \omega_n t \right)$$

$$+ c^2 \varphi_n(L) \frac{1}{\frac{c}{L} \omega_n} \int_0^t u(s) \sin \left(\frac{c}{L} \omega_n (t - s) \right) ds.$$

Then

$$y(t, x) = \sum_{n=0}^{\infty} \alpha_n(t) \, \varphi_n(x)$$

solves (RARWP). For all $t \in (0, T)$ we have

$$y(t, \cdot) \in L^2(0, L)$$

and

$$\int_0^L y(t,x)^2 \, dx = \sum_{n=0}^{\infty} (\alpha_n(t))^2 \, .$$

Moreover we have

$$y \in C((0,T), H^1(0,L)) \cap C^1((0,T), L^2(0,L)).$$

Proof. In order to obtain a series representation of the solution, we consider the eigenvalue problem

$$\varphi_{xx}(x) = -\lambda \, \varphi(x), \ x \in [0,L], \ \varphi(0) = 0, \ \varphi_x(L) = \gamma \, \varphi(L).$$

Exercise 2.6. Compute the corresponding eigenfunctions!

The corresponding eigenfunctions are

$$\varphi_n(x) = A_n \sin\left(\omega_n \frac{x}{L}\right),$$

where ω_n satisfies the equation

$$\frac{\omega_n}{L} \cos(\omega_n) = \gamma \, \sin(\omega_n).$$

This is equivalent to

$$\tan(\omega_n) = \frac{\omega_n}{\gamma L}.$$

Since $\gamma \in (-1/L, 0)$ we have $\frac{1}{\gamma L} < -1$ and thus the smallest solution $\omega_0 > 0$ is in $(\frac{1}{2} \pi, \pi)$. For $\omega_n \to \infty$ the solutions converge to the roots of the cosine functions, that is

$$\lim_{n\to\infty} \omega_n - \left(\frac{\pi}{2} + n\pi\right) = 0.$$

With the notation

$$\delta_n = \omega_n - \left(\frac{\pi}{2} + n\pi\right),$$

we get $\delta_n > 0$. The sequence $(\delta_n)_n$ converges to zero.

We obtain the eigenfunctions

$$\varphi_n(x) = A_n \sin\left(\omega_n \frac{x}{L}\right), \ n \in \{0,1,2,\ldots\} \tag{2.9}$$

with the eigenvalues

$$\lambda_n = \frac{1}{L^2}\,\omega_n^2$$

where the A_n are chosen such that

$$\int_0^L \varphi_n(x)^2\, dx = 1, \ n \in \{0, 1, 2, \ldots\}.$$

Hence the sequence $(A_n)_n$ converges to $\frac{\sqrt{2}}{\sqrt{L}}$.

The functions $((\varphi_n(x))_{n=0}^{\infty}$ form a completely orthonormal system in $L^2(0, L)$.

Exercise 2.7. Show that the $(\varphi_n)_n$ form an orthogonal system in $L^2(0, L)$.

Solution of Exercise 2.7. We get

$$\lambda_n \int_0^L \varphi_n(x)\, \varphi_m(x)\, dx$$

$$= -\int_0^L \varphi_n''(x)\, \varphi_m(x)\, dx$$

$$= \int_0^L \varphi_n'(x)\, \varphi_m'(x)\, dx - \varphi_n'(x)\, \varphi_m(x)\big|_{x=0}^{L}$$

$$= -\int_0^L \varphi_n(x)\, \varphi_m''(x)\, dx + \varphi_n(x)\, \varphi_m'(x)\big|_{x=0}^{L} - \varphi_n'(x)\, \varphi_m(x)\big|_{x=0}^{L}$$

$$= \lambda_m \int_0^L \varphi_n(x)\, \varphi_m(x)\, dx + \varphi_n(L)\, \varphi_m'(L) - \varphi_n'(L)\, \varphi_m(L)$$

$$= \lambda_m \int_0^L \varphi_n(x)\, \varphi_m(x)\, dx + \gamma\, \varphi_n(L)\, \varphi_m(L) - \gamma\, \varphi_n(L)\, \varphi_m(L)$$

$$= \lambda_m \int_0^L \varphi_n(x)\, \varphi_m(x)\, dx.$$

If $m \neq n$ this implies $\int_0^L \varphi_n(x)\, \varphi_m(x)\, dx = 0$, thus the eigenfunctions $(\varphi_n)_n$ are orthogonal. Thus Exercise 2.7 is solved.

Now we want to determine a sequence of functions $\alpha_n(t) : [0, T] \to \mathbb{R}$ such that for the solutions of (*RARWP*) we have the following series representations:

$$y(t, x) = \sum_{n=0}^{\infty} \alpha_n(t)\, \varphi_n(x). \tag{2.10}$$

For all $n \in \{0, 1, 2, \ldots\}$ we have

$$\int_0^L y_{tt}(t, x)\, \varphi_n(x)\, dx$$

$$= c^2 \int_0^L y_{xx}(t, x)\, \varphi_n(x)\, dx$$

$$= -c^2 \int_0^L y_x(t, x)\, (\varphi_n)_x(x)\, dx + c^2\, y_x(t, x)\, \varphi_n(x)|_{x=0}^L$$

$$= -c^2 \int_0^L y_x(t, x)\, (\varphi_n)_x(x)\, dx + c^2\, y_x(t, L)\, \varphi_n(L)$$

$$= c^2 \int_0^L y(t, x)\, (\varphi_n)_{xx}(x)\, dx - c^2\, y(t, x)\, (\varphi_n)_x(x)|_{x=0}^L + c^2\, y_x(t, L)\, \varphi_n(L)$$

$$= c^2 \int_0^L y(t, x)\, (\varphi_n)_{xx}(x)\, dx - c^2\, y(t, L)\, (\varphi_n)_x(L) + c^2\, y_x(t, L)\, \varphi_n(L)$$

$$= -c^2 \lambda_n \int_0^L y(t, x)\, \varphi_n(x)\, dx + c^2\, [y_x(t, L)\, \varphi_n(L) - y(t, L)\, (\varphi_n)_x(L)]$$

$$= -c^2 \lambda_n \int_0^L y(t, x)\, \varphi_n(x)\, dx + c^2\, [y_x(t, L)\, \varphi_n(L) - \gamma\, y(t, L)\, \varphi_n(L)]$$

$$= -c^2 \lambda_n \int_0^L y(t, x)\, \varphi_n(x)\, dx + c^2\, u(t)\, \varphi_n(L).$$

These equations yield the sequence of differential equations

$$\alpha_n''(t) = -\frac{c^2}{L^2} \omega_n^2 \alpha_n(t) + c^2\, \varphi_n(L)\, u(t)$$

with the initial conditions

$$\alpha_n(0) = \int_0^L y_0(x)\, \varphi_n(x)\, dx =: \alpha_n^0, \quad \alpha_n'(0) = \int_0^L y_1(x)\, \varphi_n(x)\, dx =: \alpha_n^1.$$

The solutions are

$$\alpha_n(t) = \alpha_n^0 \cos\left(\frac{c}{L} \omega_n t\right) + \alpha_n^1 \frac{1}{\frac{c}{L} \omega_n} \sin\left(\frac{a}{L} \omega_n t\right)$$

$$+ c^2\, \varphi_n(L) \frac{1}{\frac{c}{L} \omega_n} \int_0^t u(s) \sin\left(\frac{c}{L} \omega_n(t - s)\right) ds.$$

With the functions $\alpha_n(t)$ the solution $y(t, x)$ of (RARWP) has the series representation (2.10) and Theorem 2.6 is proved.□

Now we come to the question of uniqueness. Here we only consider the case of regular initial and control data.

Theorem 2.7 (Uniqueness in the regular case). *Let $y_0 \in H^2(0, L)$ with $y_0(0) = 0$, $y_1 \in H^1(0, L)$ and $u \in H^1(0, T)$ with $u(0) = \partial_x(y_0(L)) - \gamma\, y_0(L)$ be given. Assume that $\gamma \in (-1/L, 0)$. Then the solution of the initial boundary value problem (RARWP) is uniquely determined in $H^1((0, L) \times (0, T))$.*

In particular for $y_0 = y_1 = 0$ the unique solution of (RARWP) is $y(t, x) = 0$.

Proof. Let z_1 and z_2 denote solutions of (RARWP). We consider the difference

$$y = z_2 - z_1.$$

Then y is a solution of the homogeneous initial boundary value problem

$$
\begin{cases}
y(0, x) & = 0, & x \in (0, L) \\
y_t(0, x) & = 0, & x \in (0, L) \\
y_{tt}(t, x) & = c^2\, y_{xx}(t, x), & (t, x) \in (0, T) \times (0, L) \\
y(t, 0) & = 0, & t \in (0, T) \\
y_x(t, L) - \gamma y(t, L) & = 0, & t \in (0, T).
\end{cases}
$$

For $t > 0$ we have $y_x^2(t, \cdot), y_t^2(t, \cdot) \in H^1(0, L)$. Define the energy

$$E(t) = \frac{1}{2} \int_0^L y_x(t, x)^2 + \frac{1}{c^2}\, (y_t(t, x))^2 \; dx - \frac{\gamma}{2}\, (y(t, L))^2 \,.$$

Then we have

$$
\begin{aligned}
E(t) - E(0) &= \frac{1}{2} \int_0^L \left[y_x^2(\tau, x) + \frac{1}{c^2} y_t(\tau, x)^2 \right] |_{\tau=0}^t \, dx - \frac{\gamma}{2}\, (y(t, L))^2 \\
&= \int_0^L \int_0^t y_x(\tau, x)\, y_{xt}(\tau, x) + \frac{1}{c^2}\, y_t(\tau, x)\, y_{tt}(\tau, x)\, d\tau\, dx \\
&\quad - \frac{\gamma}{2}\, (y(t, L))^2 \,. \\
&= \int_0^L \int_0^t y_x(\tau, x)\, y_{xt}(\tau, x) + y_t(\tau, x)\, y_{xx}(\tau, x)\, d\tau\, dx \\
&\quad - \frac{\gamma}{2}\, (y(t, L))^2 \,.
\end{aligned}
$$

Hence we have

$$
\begin{aligned}
E(t) - E(0) &= \int_0^L \int_0^t y_x(\tau, x)\, y_{xt}(\tau, x) - y_{tx}(\tau, x)\, y_x(\tau, x)\, d\tau\, dx \\
&\quad + \int_0^t y_t(\tau, x)\, y_x(\tau, x)|_{x=0}^L\, d\tau - \frac{\gamma}{2}\, (y(t, L))^2 \\
&= 0 + \int_0^t y_t(\tau, L)\, y_x(\tau, L) - y_t(\tau, 0)\, y_x(\tau, 0)\, d\tau - \frac{\gamma}{2}\, (y(t, L))^2
\end{aligned}
$$

$$= 0 + \int_0^t y_t(\tau, L) \, y_x(\tau, L) \, d\tau - \frac{\gamma}{2} \, (y(t, L))^2$$

$$= 0 + \int_0^t \gamma \, y_t(\tau, L) \, y(\tau, L) \, d\tau - \frac{\gamma}{2} \, (y(t, L))^2$$

$$= \int_0^t \gamma \, \partial_t \frac{1}{2} \, (y(\tau, L))^2 \, d\tau - \frac{\gamma}{2} \, (y(t, L))^2$$

$$= \frac{\gamma}{2} \left[(y(t, L))^2 - (y(0, L))^2 \right] - \frac{\gamma}{2} \, (y(t, L))^2$$

$$= 0.$$

This implies

$$E(t) = E(0) = 0.$$

Thus we have $y(t, x) = 0$ and Theorem 2.7 is proved. $\square$

Remark 2.8. In the proof of Theorem 2.7 we have shown that $u(t) = 0 = \partial_x (y_0(L)) - \gamma \, y_0(L)$ implies that the energy $E(t)$ remains constant, that is it is conserved.

Remark 2.9. The solution of (*RARWP*) can be interpreted as a linear combination of a solution of (*DARWP*) and a solution of (*NAWRP*) with the corresponding initial conditions. With a control u in $L^2(0, T)$ the solution of (*DARWP*) only has the regularity

$$C((0, T), L^2(0, L)) \cap C^1((0, T), H^{-1}(0, L)).$$

Thus a linear combination of this less regular solution with the more regular solution of (*NARWP*) yields a less regular solution. In order to get the regularity of (*NARWP*), that is

$$C((0, T), H^1(0, L)) \cap C^1((0, T), L^2(0, L))$$

for the linear combinations, we need more regular controls $u \in H^1(0, T)$.

Chapter 3
Exact Controllability

The question of exact controllability (see [44, 49]) is: Which states can be reached exactly at given control time T with a given set of control functions starting at time zero with an initial state from a prescribed set?

If the state at the terminal time T is prescribed a priori, this leads to end conditions that have the same form as the initial conditions. These condition can be prescribed as equality constraints in optimal control problems.

For the systems (*DARWP*), (*NARWP*), and (*RARWP*), the end conditions have a similar form as the initial conditions, namely

$$\begin{cases} y(T,x) = z_0(x), \, x \in (0,L), \\ y_t(T,x) = z_1(x), \, x \in (0,L). \end{cases}$$

For reversible systems, the question whether an initial state (y_0, y_1) can be controlled in such a way that at the time T the state $(0, 0)$ is reached exactly is equivalent to the question, whether the system can be controlled from $(0,0)$ to (z_0, z_1). Since the system is also linear, the question whether the system can be controlled from the initial state (y_0, y_1) to the terminal state (z_0, z_1) can be reduced to the question, whether the system can be controlled from a given initial state to $(0, 0)$. On account of the finite speed of the propagation of information in hyperbolic systems, a necessary condition for exact controllability for systems governed by hyperbolic systems and in particular by the wave equation is that T is sufficiently large.

© The Author(s) 2015
M. Gugat, *Optimal Boundary Control and Boundary Stabilization of Hyperbolic Systems*, SpringerBriefs in Electrical and Computer Engineering,
DOI 10.1007/978-3-319-18890-4_3

3.1   Dirichlet boundary control

Let a space interval $[0, L]$, a time interval $[0, T]$ and a wave speed $c > 0$ be given. We consider the initial boundary value problem for the wave equation

$$y_{tt}(t, x) = c^2 y_{xx}(t, x), \quad (t, x) \in [0, T] \times [0, L], \tag{3.1}$$

with the initial conditions

$$y(0, x) = 0, \quad x \in [0, L], \tag{3.2}$$

$$y_t(0, x) = 0, \quad x \in [0, L] \tag{3.3}$$

and the Dirichlet boundary conditions

$$y(t, 0) = f_1(t), \quad t \in [0, T], \tag{3.4}$$

$$y(t, L) = f_2(t), \quad t \in [0, T]. \tag{3.5}$$

To prescribe the desired target state we use the end conditions

$$y(T, x) = y_0(x), \quad x \in [0, L], \tag{3.6}$$

$$y_t(T, x) = y_1(x), \quad x \in [0, L]. \tag{3.7}$$

Here y_0 is in $L^2(0, L)$ and y_1 possesses an antiderivative in the space $L^2(0, L)$. Thus we have $Y_1(x) = \int_0^x y_1(z) \, dz \in L^2(0, L)$, that is $y_1 \in H^{-1}(0, L)$.

Theorem 3.1 (Exact controllability, see [33]). *Let $p \in [2, \infty]$ and $T \geq L/c$ be given. The initial boundary value problem (3.1)–(3.5) has a weak solution that satisfies the end conditions (3.6)–(3.7) with $f_1, f_2 \in L^p(0, T)$, if and only if y_0 and y_1 satisfy the following conditions: $y_0 \in L^p(0, L)$ and $Y_1 \in L^p(0, L)$, with $Y_1(x) = \int_0^x y_1(z) \, dz$. Thus we have $y_1 \in W^{-1,p}(0, L)$.*

Proof. For the proof we will use the fact that the set of admissible controls that generate states that satisfy the prescribed end conditions can be characterized as the solution set of a sequence of trigonometric moment equations. This is a classical approach to exact controllability, see [40, 49]. To derive these equations, we need the series representation of the solution of $(DARWP)$.

   Step 1: Series representation of the solution of $(DARWP)$**.** We use

$$\varphi \in C^2([0, L])$$

as a test function. From (3.1) we get

$$\int_0^L y_{tt}(t, x) \, \varphi(x) \, dx = \int_0^L c^2 \, y_{xx}(t, x) \, \varphi(x) \, dx.$$

With integration by parts we get from the right-hand side

$$\int_0^L y_{xx}(t,\,x)\,\varphi(x)\,dx = -\int_0^L y_x(t,\,x)\,\varphi_x(x)\,dx + y_x(t,\,x)\varphi(x)|_{x=0}^L.$$

The boundary conditions (3.4), (3.5) and another integration by part yield

$$\int_0^L y_{xx}(t,\,x)\,\varphi(x)\,dx = \int_0^L y(t,\,x)\varphi_{xx}(x)\,dx - f_2(t)\varphi_x(L) + f_1(t)\varphi_x(0)$$
$$+y_x(t,\,x)\varphi(x)|_{x=0}^L.$$

For test functions φ that satisfy the homogeneous boundary conditions $\varphi(0) = 0 = \varphi(L)$, this implies

$$\int_0^L y_{tt}(t,\,x)\,\varphi(x)\,dx = \int_0^L c^2\,y(t,\,x)\,\varphi_{xx}(x)\,dx + c^2(f_1(t)\varphi_x(0) - f_2(t)\varphi_x(L)). \qquad (3.8)$$

Now we consider the eigenvalue problem

$$c^2\varphi_{xx}(x) + \lambda\varphi(x) = 0,\ x \in [0, L],\ \varphi(0) = 0 = \varphi(L).$$

The corresponding eigenfunctions are

$$\varphi_j(x) = (\sqrt{2}/\sqrt{L})\,\sin(j\pi x/L)$$

with the normalization

$$\int_0^L \varphi_j^2(x)\,dx = 1,\ \ j \in \mathbb{N}$$

and the eigenvalues $\lambda_j = (cj\pi/L)^2$, $j \in \mathbb{N}$. For all $j \in \mathbb{N}$ we have

$$\varphi_j'(0) = (\sqrt{2}/\sqrt{L})(j\pi/L),$$
$$\varphi_j'(L) = (-1)^j(\sqrt{2}/\sqrt{L})(j\pi/L) = (-1)^j\varphi_j'(0).$$

The functions $(\varphi_j)_{j\in\mathbb{N}}$ form a complete orthonormal system in $L^2(0, L)$. On account of the special form of the eigenfunctions, the expansion with respect to the φ_j yields a Fourier-sine series. Now we consider the expansion of $y(t, \cdot)$ as a series of eigenfunctions for $t \in (0, T)$:

$$y(t,\,x) = \sum_{j=1}^{\infty} \alpha_j(t)\,\varphi_j(x).$$

With this representation of $y(t, \cdot)$ the boundary conditions (3.4)–(3.5) can only hold in a weak sense. The equation

$$\int_0^L y_{tt}(t, x)\varphi_j(x)\, dx = \frac{d^2}{dt^2} \int_0^L y(t, x)\varphi_j(x)\, dx = \alpha_j''(t)$$

and (3.8) yield the differential equation

$$\alpha_j''(t) = \int_0^L y(t, x)(-\lambda_j\varphi_j(x))\, dx + c^2 \left[f_1(t)\, \varphi_j'(0) - f_2(t)\, \varphi_j'(L)\right]$$

$$= -\lambda_j\alpha_j(t) + c^2\varphi_j'(0)\left[f_1(t) - (-1)^j f_2(t)\right], \ j \in \mathbb{N}.$$

From the initial conditions (3.2)–(3.3) we get $\alpha_j(0) = 0$, $\alpha_j'(0) = 0$ and with the ordinary differential equations

$$\alpha_j''(t) + \lambda_j\,\alpha_j(t) = c^2\,\varphi_j'(0)\left[f_1(t) - (-1)^j f_2(t)\right]$$

we get an initial value problem for α_j. For $j \in \mathbb{N}$ the solution is

$$\alpha_j(t) = \int_0^t \left[f_1(s) - (-1)^j f_2(s)\right] c\,(\sqrt{2}/\sqrt{L})\sin\left(c(\pi j/L)(t - s)\right)\, ds.$$

Hence y has the series representation

$$y(t, x) = \sum_{j=1}^{\infty} \frac{2c}{L} \int_0^t \left[f_1(s) - (-1)^j f_2(s)\right]\sin\left((c\pi j/L)(t - s)\right)\, ds \sin\left((j\pi/L)x\right)$$

and the velocity y_t is given by

$$y_t(t, x) = \sum_{j=1}^{\infty} \frac{2c^2 j\pi}{L^2} \int_0^t \left[f_1(s) - (-1)^j f_2(s)\right]\cos\left(\frac{c\pi j}{L}(t - s)\right)\, ds \sin\left(\frac{j\pi}{L}x\right).$$

Step 2: **End conditions and a trigonometric moment problem**

By inserting the series representations of y and y_t in the end conditions (3.6), (3.7), we get the equations

$$\alpha_j(T) = \int_0^L y_0(x)\varphi_j(x)\, dx, \ j \in \mathbb{N}, \tag{3.9}$$

$$\alpha_j'(T) = \int_0^L y_1(x)\varphi_j(x)\, dx, \ j \in \mathbb{N}. \tag{3.10}$$

Now (3.9) yields the trigonometric moment equations

$$\int_0^T \frac{\sqrt{2}c}{\sqrt{L}}[f_1(s) - (-1)^j f_2(s)] \sin\left(\frac{c\pi j}{L}(T-s)\right) ds = \int_0^L y_0(x)\varphi_j(x)\, dx \quad (3.11)$$

and (3.10) implies

$$\int_0^T \frac{\sqrt{2}c^2\pi j}{L^{3/2}}[f_1(s) - (-1)^j f_2(s)] \cos\left(\frac{c\pi j}{L}(T-s)\right) ds = \int_0^L y_1(x)\varphi_j(x)\, dx \tag{3.12}$$

for all $j \in \mathbb{N}$. In this way we have characterized the set of admissible controls as the solution set of a trigonometric moment problem. This is a classical approach to problems of exact controllability (see [49]).

Step 3: The minimal time interval with exact controllability

Now we consider the exact controllability on the time interval for

$$T = L/c.$$

Since this is the minimal time for which the state at time T can be influenced by both boundary controls at all points in the space interval, it is clear that this is the minimal time for which there is a chance that exact controllability holds. For given control functions f_1 and f_2 we define the sum

$$S(t) = \frac{f_1(T-t) + f_2(T-t)}{2} \tag{3.13}$$

and the difference

$$D(t) = \frac{f_1(T-t) - f_2(T-t)}{2}. \tag{3.14}$$

The trigonometric moment equations (3.11), (3.12) are equivalent to moment equations for S and D. We start with the moment problem for D:

$$\int_0^{Tc} D(t/c)(\sqrt{2}/\sqrt{L}) \sin(2\pi jt/L)\, dt = \frac{y_0^{2j}}{2}, \tag{3.15}$$

$$\int_0^{Tc} D(t/c)(\sqrt{2}/\sqrt{L}) \cos(2\pi jt/L)\, dt = \frac{L y_1^{2j}}{4c\pi j}. \tag{3.16}$$

From these equations, we get all the Fourier coefficients of the function $D(\cdot/c)$ except for the Fourier coefficient for the constant function. Hence there exists a real number r, such that for all $x \in [0, L]$ we have

$$D(x/c) = r + \sum_{j=1}^{\infty}(y_0^{2j}/2)\sqrt{(2/L)}\sin(2\pi jx/L)$$

$$- \sum_{j=1}^{\infty}[1/(2c)]\,(-y_1^{2j}L/(2\pi j))\,\sqrt{(2/L)}\cos(2\pi jx/L).$$

We define the function

$$Y_1^{\text{even}}(x) = \sum_{j=1}^{\infty} -y_1^{2j}(L/(2\pi j))\sqrt{(2/L)}\cos((2\pi j/L)x)$$

and the functions

$$y_1^{\text{odd}}(x) = \sum_{j=1}^{\infty}(y_1^{2j})\sqrt{(2/L)}\sin(2\pi jx/L),$$

$$y_0^{\text{odd}}(x) = \sum_{j=1}^{\infty}(y_0^{2j})\sqrt{(2/L)}\sin(2\pi jx/L).$$

We have

$$(Y_1^{\text{even}})'(x) = \sum_{j=1}^{\infty}(y_1^{2j})\sin(2\pi jx/L) = y_1^{\text{odd}}(x)$$

and for D for $x \in [0, L/c]$ we have the representation

$$D(x) = r + y_0^{\text{odd}}(cx)/2 - (1/(2c))\,Y_1^{\text{even}}(cx). \tag{3.17}$$

The function Y_1^{even} is symmetric on the interval $[0, L]$ with respect to the point $L/2$, that is for $x \in [-L/2, L/2]$ we have

$$Y_1^{\text{even}}(L/2 - x) = Y_1^{\text{even}}(L/2 + x).$$

The function y_0^{odd} is odd on the interval $[0, L]$ with respect to the point $L/2$, that is for $x \in [-L/2, L/2]$ we have

$$y_0^{\text{odd}}(L/2 - x) = -y_0^{\text{odd}}(L/2 + x).$$

Now we consider the moment problem for S:

$$\int_0^{Tc} S(t/c)(\sqrt{2}/\sqrt{L})\sin((2j-1)\pi t/L)\,dt = \frac{y_0^{2j-1}}{2}, \tag{3.18}$$

$$\int_0^{Tc} S(t/c)(\sqrt{2}/\sqrt{L})\cos((2j-1)\pi t/L)\,dt = \frac{L\,y_1^{2j-1}}{2(2j-1)\,\pi\,c}. \qquad (3.19)$$

We define the functions

$$y_0^{\text{even}}(x) = \sum_{j=1}^{\infty}(y_0^{2j-1})\sqrt{(2/L)}\sin((2j-1)\pi x/L),$$

$$y_1^{\text{even}}(x) = \sum_{j=1}^{\infty}(y_1^{2j-1})\sqrt{(2/L)}\sin((2j-1)\pi x/L)$$

and the function

$$Y_1^{\text{odd}}(x) = \sum_{j=1}^{\infty}-y_1^{2j-1}(L/((2j-1)\pi))\sqrt{(2/L)}\cos(((2j-1)\pi/L)x).$$

Then we have

$$(Y_1^{\text{odd}})'(x) = y_1^{\text{even}}(x).$$

For S for $x \in [0, L/c]$ we have the representation

$$S(x) = y_0^{\text{even}}(cx)/2 - (1/(2c))\,Y_1^{\text{odd}}(cx). \qquad (3.20)$$

Hence for the controls f_1, f_2 that steer the system at the time $T = L/c$ to the desired target state we have the equations

$$f_1(T - t) = S(t) + D(t) \qquad (3.21)$$

$$= r + y_0(ct)/2 - (1/(2c))\,Y_1(ct)$$

and

$$f_2(T - t) = S(t) - D(t) \qquad (3.22)$$

$$= -r + (y_0^{\text{even}}(ct) - y_0^{\text{odd}}(ct))/2 + (1/(2c))(Y_1^{\text{even}}(ct) - Y_1^{\text{odd}}(ct))$$

with a real number r.

This representation of f_1 and f_2 implies that if y_0 and Y_1 are in the space $L^p(0, L)$, the control functions f_1 and f_2 are elements of the space $L^p(0, L/c)$.

On the other hand for f_1 and $f_2 \in L^p(0, L/c)$, the definitions of S and D imply that we also have S and $D \in L^p(0, L/c)$, which in turn implies that $y_0^{\text{odd}}, Y_1^{\text{even}}, y_0^{\text{even}}, Y_1^{\text{odd}} \in L^p(0, L)$. This implies $y_0, Y_1 \in L^p(0, L)$.

Thus we have shown the assertion of Theorem 3.1 for the minimal control time $T = L/c$.

For the minimal control time, the admissible controls f_1 and $f_2 \in L^p(0, T)$, that steer the system to a position of rest, are determined uniquely up to the constant r in (3.21), (3.22). This is a consequence of the fact that the moment problem for S has a unique solution in $L^2(0, T)$ and the moment problem for D determines the function D up to an additive constant.

Step 4: Exact controllability for larger time intervals

Now we consider the case $T > L/c$ and reduce it to the case of the minimal time where exact controllability holds that we have considered in Step 3 of this proof.

Let y_0 and $Y_1 \in L^p(0, L)$ be given. Then from Step 3 we get admissible controls f_1 and f_2 in $L^p(0, L/c)$, that steer the system to a position of rest at the time L/c. We extend f_1 and f_2 by zero to functions in $L^p(0, T)$, that is we put $f_1(t) = f_2(t) = 0$ for $t > L/c$. Since the system is in a position of rest at the time L/c, with these controls it remains at rest and therefore satisfies the end conditions at the terminal time T. Thus we have constructed admissible controls f_1 and f_2 in $L^p(0, T)$.

Now we show that also the converse holds. For this purpose we first describe a general construction of a function that is defined on the minimal time interval $[0, L/c]$ starting from a function that is defined on $[0, T]$. In Step 5 we will apply this construction to S and D.

Let admissible control functions f_1 and f_2 in $L^p(0, T)$ be given. Since $T > L/c$, we can choose a natural number k, such that $kL/c \leq T < (k+1)L/c$. Let the function $\varphi(s)$ be an element of the set

$$\{\sin((c\pi j/L)s), \; \cos((c\pi j/L)s) \; \text{ with } j \in \mathbb{N}, j \text{ odd}\}.$$

Then we have $\varphi(s + L/c) = -\varphi(s)$. For all $v \in L^2(0, T)$ we have

$$\int_0^T v(s)\varphi(s)\,ds = \left[\sum_{j=0}^{k-1} \int_{jL/c}^{(j+1)L/c} v(s)\varphi(s)\,ds\right] + \int_{kL/c}^T v(s)\varphi(s)\,ds$$

$$= \left[\sum_{j=0}^{k-1} \int_0^{L/c} (-1)^j v(s + jL/c)\varphi(s)\,ds\right]$$

$$+ \int_0^{T-kL/c} (-1)^k v(s + kL/c)\varphi(s)\,ds$$

$$= \sum_{j=0}^k \int_0^{T-kL/c} (-1)^j v(s + jL/c)\varphi(s)\,ds$$

$$+ \sum_{j=0}^{k-1} \int_{T-kL/c}^{L/c} (-1)^j v(s + jL/c)\varphi(s)\,ds$$

$$= \int_0^{T-kL/c} \left[\sum_{j=0}^{k} (-1)^j v(s+jL/c) \right] \varphi(s)\, ds$$

$$+ \int_{T-kL/c}^{L/c} \left[\sum_{j=0}^{k-1} (-1)^j v(s+jL/c) \right] \varphi(s)\, ds$$

We define

$$\hat{v}(t) = \sum_{j=0}^{k} (-1)^j v(t+jL/c) \tag{3.23}$$

for $t \in (0, T - kL/c)$ and

$$\hat{v}(t) = \sum_{j=0}^{k-1} (-1)^j v(t+jL/c) \tag{3.24}$$

for $t \in (T - kL/c, L/c)$. Then we have

$$\int_0^{L/c} \hat{v}(s)\varphi(s)\, ds = \int_0^{T} v(s)\varphi(s)\, ds.$$

This allows us to complete the proof in Step 5.

Step 5: Completion of the proof of Theorem 3.1

Let $f_1, f_2 \in L^p(0, T)$ be given. Let S and D be defined by (3.13) and (3.14). Then we have $S, D \in L^p(0, T)$. We define $\hat{S}, \hat{D}$ as

$$\hat{S}(t) = \sum_{j=0}^{k} (-1)^j S(t+jL/c), \tag{3.25}$$

$$\hat{D}(t) = \sum_{j=0}^{k} (-1)^j D(t+jL/c), \tag{3.26}$$

for $t \in (0, T - kL/c)$ and

$$\hat{S}(t) = \sum_{j=0}^{k-1} (-1)^j S(t+jL/c), \tag{3.27}$$

$$\hat{D}(t) = \sum_{j=0}^{k-1} (-1)^j D(t+jL/c), \tag{3.28}$$

for $t \in (T - kL/c, L/c)$ with $\hat{r} \in \mathbb{R}$. Then we have $\hat{S}, \hat{D} \in L^p(0, L/c)$.

On account of Step 4 the definition of $\hat{S}$ implies that $\hat{S}$ satisfies the moment equations (3.18), (3.19) with integrals on the interval $[0, L]$. Thus from Step 3 we get

$$\hat{S}(x) = y_0^{\text{even}}(cx)/2 - (1/(2c)) Y_1^{\text{odd}}(cx). \tag{3.29}$$

Analogously we conclude that $\hat{D}$ satisfies the moment equations (3.15), (3.16) with integrals on the interval $[0, L]$. Thus from Step 3 we get

$$\hat{D}(x) = \hat{r} + y_0^{\text{odd}}(cx)/2 - (1/(2c)) Y_1^{\text{even}}(cx) \tag{3.30}$$

From the definition we have $\hat{S}$, $\hat{D} \in L^p(0, L/c)$, and as in Step 3 by (3.29), (3.30) this implies y_0, $Y_1 \in L^p(0, L)$. Thus we have proved Theorem 3.1.$\square$

Exercise 3.1. Let $L = 1$ and $c = 1$. Then we have $t_0 = 1$.

a) Let $y_0(x) = -2$ and $y_1(x) = 0$. Determine all pairs of Dirichlet boundary controls (f_1, f_2) that control the system to a position of rest at the time t_0.

b) Determine again for $T = t_0$ such a pair, for which

$$\int_0^T f_1(s)^2(s) + f_2(s)^2(s) \, ds$$

is minimal.

c) Determine for $T = 3\, t_0$ such a pair, for which

$$\int_0^T f_1(s)^2(s) + f_2(s)^2(s) \, ds$$

is minimal.

d) Determine for $T = 3\, t_0$ such a pair, for which

$$\max\{ \|f_1\|_{L^\infty(0,T)}, \ \|f_2\|_{L^\infty(0,T)}\}$$

is minimal.

Solution of Exercise 3.1.

a) *We have $Y_1(x) = 0$. As in Step 3 of the proof of Theorem 3.1 we get f_1 and f_2. For this purpose we compute $y_0^{\text{odd}}(x) = 0$, $y_0^{\text{even}}(x) = y_0$. The functions y_0^{odd}, y_0^{even} are uniquely determined by the decomposition ($x \in [-L/2,\ L/2]$)*

$$y_0(x+L/2) = y_0^{\text{odd}}(x + L/2) + y_0^{\text{even}}(x + L/2)$$

$$= \frac{1}{2}[y_0(x + L/2) + y_0(L/2 - x)] + \frac{1}{2}[y_0(x + L/2) - y_0(L/2 - x)].$$

We have

$$y_0^{odd}(L/2 - x) = -y_0^{odd}(x + L/2), \ y_0^{even}(L/2 - x) = y_0^{even}(L/2 + x).$$

Let r be a real number. For $t \in (0, t_0)$ we get the controls

$$f_1(t) = r - 1, \ f_2(t) = -r - 1.$$

Then we have for $t = 1$: $y(t, x) = 0$ and $y_t(t, x) = 0$ for all $x \in (0, L)$.
b) *We obtain the optimal controls in the sense of b) for $r = 0$. We get the optimal controls*

$$f_1(t) = -1, \ f_2(t) = -1.$$

c) *For $j \in \{0, 1, 2\}$ and $t \in [0, t_0)$ we get*

$$f_1(t + jt_0) = f_2(t + jt_0) = (-1)^{j+1}/3.$$

d) *We get the same controls as in c).*

3.2 Neumann boundary control

Now we study the exact controllability with Neumann boundary control. Let a space interval $[0, L]$, a time interval $[0, T]$ and the wave speed $c = 1$ be given. We consider the initial boundary value problem (*NARWP*) for the wave equation. To prescribe the desired terminal state we use the end conditions

$$y(T, x) = 0, \quad x \in [0, L], \tag{3.31}$$

$$y_t(T, x) = 0, \quad x \in [0, L]. \tag{3.32}$$

For our analysis we start with a traveling waves solution of (*NARWP*) that is of the same type as the traveling waves solution of (*DARWP*) that we have seen in Theorem 2.1. These solutions allow us to characterize the successful controls in a form where no moment equations appear.

3.2.1 A traveling waves solution

We are looking for a solution of (*NARWP*) that has the form

$$y(t, x) = \alpha(x + t) + \beta(x - t),$$

where α and β are determined by the initial and boundary data. For the sake of simplicity we put $L = 1$ and $a = 1$. For $t \in (0, 1)$ the initial conditions imply the equations

$$\alpha(t) = \frac{1}{2}\left(y_0(t) + \int_0^t y_1(s)\, ds\right) + C_0, \tag{3.33}$$

$$\beta(t) = \frac{1}{2}\left(y_0(t) - \int_0^t y_1(s)\, ds\right) - C_0 \tag{3.34}$$

with a constant $C_0 \in \mathbb{R}$. For $t > 0$ the boundary condition $y(t, 0) = 0$ implies the equation $0 = \alpha(t) + \beta(-t)$, hence we have for all $s < 0$

$$\beta(s) = -\alpha(-s). \tag{3.35}$$

For the partial derivatives of y we get

$$y_x(t, x) = \alpha'(t + x) + \beta'(x - t), \tag{3.36}$$
$$y_t(t, x) = \alpha'(t + x) - \beta'(x - t). \tag{3.37}$$

Thus the boundary condition at $x = 1$ implies

$$y_x(t, 1) = \alpha'(1 + t) + \beta'(1 - t) = u(t).$$

For the derivative α' this implies the equation

$$\alpha'(1 + t) = u(t) - \beta'(1 - t).$$

By integration from 0 to t this yields

$$\alpha(1 + t) - \alpha(1) = \beta(1 - t) - \beta(1) + \int_0^t u(s)\, ds.$$

Thus we have

$$\alpha(t + 1) = \beta(1 - t) + \int_0^t u(s)\, ds + [\alpha(1) - \beta(1)].$$

With (3.33) and (3.34) this implies

$$\alpha(1) - \beta(1) = \int_0^1 y_1(s)\, ds + 2\, C_0.$$

With the choice

$$C_0 = -\frac{1}{2}\int_0^1 y_1(s)\, ds$$

we get $\alpha(1) - \beta(1) = 0$ and thus

$$\alpha(t+1) = \beta(1-t) + \int_0^t u(s)\,ds. \tag{3.38}$$

With the values of α for $t \in (0,1)$ from equation (3.33), (3.35) yields $\beta|_{(-1,0)}$. For $t \in (0,1)$ we get the values of β from (3.34).

With the known values of $\beta|_{(-1,1)}$, equation (3.38) yields the values of α on $(1,3)$. On account of (3.35) for $t \geq 1$ we have

$$\alpha(t+1) = -\alpha(t-1) + \int_0^t u(s)\,ds$$

which is equivalent to the following equation for $t \geq 0$:

$$\alpha(t+2) = -\alpha(t) + \int_0^{t+1} u(s)\,ds \tag{3.39}$$

Equation (3.39) allows us to determine the values of α recursively: We start with $\alpha|_{(1,3)}$, and with the values of the control $u(t)$ we get $\alpha|_{(3,5)}$. Then (3.39) yields $\alpha|_{(7,9)}$ and we can continue the construction.

To describe the construction completely in terms of α without using β, we extend the domain of α by the interval $(-1,0)$.

We define $\alpha|_{(-1,0)}$ with (3.35) for $s \in (0,1)$.

This implies $\alpha(t) = -\beta(-t)$ for $t \in (-1,0)$ with the values of $\beta|_{(0,1)}$ from (3.34). Then for $t \in (-1,0)$ equation (3.38) yields

$$\alpha(t+2) = -\alpha(t) + \int_0^{t+1} u(s)\,ds.$$

In the following lemma we summarize the construction of α.

Lemma 3.1. *Let $y_0 \in H^1(0,1)$ with $y_0(0) = 0$ and $y_1 \in L^2(0,1)$ be given. Define*

$$C_0 = -\frac{1}{2}\int_0^1 y_1(s)\,ds. \tag{3.40}$$

We define $\alpha \in L^2(-1,1)$ as

$$\alpha(t) = \frac{1}{2}\left(-y_0(-t) + \int_0^{-t} y_1(s)\,ds\right) + C_0 \text{ for } t \in (-1,0), \tag{3.41}$$

$$\alpha(t) = \frac{1}{2}\left(y_0(t) + \int_0^t y_1(s)\,ds\right) + C_0 \text{ for } t \in [0,1). \tag{3.42}$$

Let a natural number $K \in \{1, 2, 3, \ldots\}$, $T = 2K$ *and a control* $u \in L^2(0, T)$ *be given. For* $k \in \{1, 2, 3, \ldots\}$ *and* $t \in (-1, 1)$ *with* $t < T + 1 - 2k$ *we define the values of* α *recursively by*

$$\alpha(t + 2k) = -\alpha(t + 2(k - 1)) + \int_0^{t+2k-1} u(s)\, ds. \tag{3.43}$$

Then α *is well defined on the interval* $(-1, T + 1)$ *and*

$$\alpha \in H^1(-1, T + 1).$$

Proof. The construction implies

$$\alpha|_{(k-1,k)} \in H^1(k - 1, k)$$

for all $k \in \{0, 1, 2, 3, \ldots\}$. To show that $\alpha \in H^1(-1, T + 1)$, it suffices to show that α is continuous. Since $\alpha(0+) = \alpha(0-) = C_0$, α is continuous at $t = 0$. On account of the definition of α by (3.41), (3.42) this implies that α is continuous on $(-1, 1)$. At $t = 1$, on account of the definition of C_0 in (3.40) we have $\alpha(1+) = \alpha(1-) = \frac{1}{2}y_0(1)$. Therefore α is continuous at $t = 1$. This implies that α is continuous on the interval $(-1, 3)$.

Now we proceed inductively. Let $k \in \{1, 2, 3, \ldots\}$. We assume that α is continuous on $(-1, 1 + 2k)$. Then we have

$$\alpha((-1 + 2k)-) = \alpha((-1 + 2k)+).$$

Equation (3.43) implies

$$\alpha((1 + 2k)-) = -\alpha((1 + 2(k - 1))-) + \int_0^{2k} u(s)\, ds$$

$$= -\alpha((-1 + 2k)-) + \int_0^{2k} u(s)\, ds$$

$$= -\alpha((-1 + 2k)+) + \int_0^{2k} u(s)\, ds$$

$$= \alpha((-1 + 2(k + 1))+)$$

$$= \alpha((1 + 2k)+).$$

On account of the definition of α by (3.41), (3.42) by induction this implies that α is continuous on $(-1, 1 + 2(k + 1))$ as long as $k + 1 \leq K$. Thus we have proved Lemma 3.1. $\square$

Now we use α from Lemma 3.1 to get a representation of the solution of (*NARWP*).

Theorem 3.2. *Let $y_0 \in H^1(0, 1)$ with $y_0(0) = 0$ and $y_1 \in L^2(0, 1)$ be given.*
Let $K \in \{1, 2, 3, \ldots\}$, $T = 2K$ and a control $u \in L^2(0, T)$ be given.
Let α be defined as in Lemma 3.1. Then the solution $y(t, x)$ of (NARWP) for $c = 1$
is given by

$$y(t, x) = \alpha(t + x) - \alpha(t - x), \quad (t, x) \in (0, T) \times (0, 1). \tag{3.44}$$

For $x \in (0, 1)$, at the time T we have

$$y(T, x) = \alpha(T + x) - \alpha(T - x) \tag{3.45}$$

$$= (-1)^K y_0(x) - \sum_{k=1}^{K} (-1)^k \int_{1-x}^{1+x} u(2(K - k) + s) \, ds \tag{3.46}$$

and the spatial partial derivative is

$$y_x(T, x) = (-1)^K \partial_x y_0(x) \tag{3.47}$$

$$- \sum_{k=0}^{K-1} (-1)^{K-k} \left[u(2k + (1 + x)) + u(2k + (1 - x)) \right]. \tag{3.48}$$

The corresponding velocity is

$$y_t(T, x) = (-1)^K y_1(x) \tag{3.49}$$

$$- \sum_{k=0}^{K-1} (-1)^{K-k} \left[u(2k + (1 + x)) - u(2k + (1 - x)) \right]. \tag{3.50}$$

Proof. The construction of α implies that y from (3.44) is a solution of (NARWP).
For $t = 2k_0$ with $k_0 \in \{1, 2, 3, \ldots, K\}$, (3.43) implies

$$y(2k_0, x) = \alpha(2k_0 + x) - \alpha(2k_0 - x)$$

$$= -[\alpha(2(k_0 - 1) + x) - \alpha(2(k_0 - 1) - x)] + \int_{2k_0 - 1 - x}^{2k_0 - 1 + x} u(s) \, ds$$

$$= -y(2(k_0 - 1), x) + \int_{2(k_0 - 1) + 1 - x}^{2(k_0 - 1) + 1 + x} u(s) \, ds.$$

For $k_0 \geq 2$ this implies

$$y(2(k_0 - 1), x) = -y(2(k_0 - 2), x) + \int_{2(k_0 - 2) + 1 - x}^{2(k_0 - 2) + 1 + x} u(s) \, ds$$

and

$$y(2k_0, x) = y(2(k_0 - 2), x) - \int_{2(k_0 - 2) + 1 - x}^{2(k_0 - 2) + 1 + x} u(s) \, ds + \int_{2(k_0 - 1) + 1 - x}^{2(k_0 - 1) + 1 + x} u(s) \, ds.$$

For $T = 2K$ we get by induction

$$y(2K, x) = (-1)^K y_0(x) - \sum_{k=1}^{K} (-1)^k \int_{2(K-k)+1-x}^{2(K-k)+1+x} u(s) \, ds$$

and hence (3.46) holds. By partial differentiation and an index transformation in the sum this implies (3.47)–(3.48).

Now we compute the velocity $y_t(T, x)$.

We have $y_t(t, x) = \alpha'(t + x) - \alpha'(t - x)$. From (3.43) we get the recursion

$$\alpha'(t + 2k) = -\alpha'(t + 2(k-1)) + u(t + 2k - 1). \qquad (3.51)$$

For $t = 2k_0$ with $k_0 \in \{1, 2, 3, \ldots, K\}$ this implies

$$
\begin{aligned}
y_t(2k_0, x) &= \alpha'(2k_0 + x) - \alpha'(2k_0 - x) \\
&= -\big[\alpha'(2(k_0 - 1) + x) - \alpha'(2(k_0 - 1) - x)\big] \\
&\quad + [u(2k_0 - 1 + x) - u(2k_0 - 1 - x)] \\
&= -y_t(2(k_0 - 1), x) + [u(2k_0 - 1 + x) - u(2k_0 - 1 - x)].
\end{aligned}
$$

Hence for $k_0 \geq 2$ we have

$$y_t(2(k_0 - 1), x) = -y_t(2(k_0 - 2), x) + [u(2(k_0 - 2) + 1 + x) - u(2(k_0 - 2) + 1 - x)].$$

and

$$
\begin{aligned}
y_t(2k_0, x) &= y_t(2(k_0 - 2), x) - [u(2(k_0 - 2) + 1 + x) - u(2(k_0 - 2) + 1 - x)] \\
&\quad + [u(2k_0 - 1 + x) - u(2k_0 - 1 - x)].
\end{aligned}
$$

For $T = 2K$ by induction we get

$$
\begin{aligned}
y_t(2K, x) &= (-1)^K y_1(x) \\
&\quad - \sum_{k=1}^{K} (-1)^k \, [u(2(K - k) + 1 + x) - u(2(K - k) + 1 - x)]
\end{aligned}
$$

and thus (3.49)–(3.50). Hence Theorem 3.2 is proved. $\square$

3.2.2 Exact Controllability with Neumann boundary control

In this section we give a result about the exact controllability of (*NARWP*) that also follows from the result about optimal control in Theorem 2.1 in [26] (see also Section 4.2).

Theorem 3.3 (Exact controllability). *Let $T \geq 2$, $L = 1$, $a = 1$ be given. The initial boundary value problem (NARWP) has a weak solution that satisfies the end conditions*

$$y(T, x) = 0, \ y_t(T, x) = 0, \ x \in (0, 1)$$

with $u \in L^2(0, T)$, if and only if y_0 and y_1 satisfy the following conditions: $y_0 \in H^1(0, L)$, $y_0(0) = 0$ and $y_1 \in L^2(0, L)$.

Exercise 3.2. Use Theorem 3.2 to prove Theorem 3.3 for the case $T = 2K$.

For this purpose, construct a control u with the right regularity that steers a given initial state (y_0, y_1) to a position of rest at the minimal control time $T = 2$.

Find the control by inserting in (3.47)–(3.48) and (3.49)–(3.50) the desired terminal state and solve the resulting system of linear equations for the control values. Note that $y_x(T, x) = 0$ for $x \in (0, 1)$ almost everywhere and $y(T, 0) = 0$ imply on account of the regularity of the solution, that $y(T, x) = 0$ for $x \in (0, 1)$ almost everywhere.

To prove the converse for $T = 2K$, solve (3.47)–(3.48) to get y_0 and (3.49)–(3.50) to get y_1 by inserting the known terminal state which is the position of rest. Now use the regularity of the controls to deduce the regularity of the initial state.

Remark 3.1. For the minimal control time $T = 2$ the control that steers the system to a position of rest is *uniquely* determined by the initial state. Thus also the corresponding state y is determined uniquely! For initial data that satisfies the regularity assumptions of Theorem 3.3, the state is the same as for exact Dirichlet boundary control.

3.3 Robin boundary control

Now we analyse the exact controllability for Robin boundary controls. Since ROBIN-boundary controls are a linear combination of Dirichlet- and Neumann boundary controls, the strategy to get an exact controllability result is to use the exact Neumann-boundary control (that yields a more regular state than the Dirichlet control) to get the Robin-boundary control. This yields the following theorem.

Theorem 3.4 (Exact controllability). *Let $T \geq 2$, $L = 1$, $a = 1$ and $\gamma < 0$ be given. If $y_0 \in H^1(0, L)$, $y_0(0) = 0$ and $y_1 \in L^2(0, L)$, there is a control $u \in L^2(0, T)$, such that the initial boundary value problem (RARWP) has a weak solution that satisfies the end conditions*

$$y(T, x) = 0, \ y_t(T, x) = 0, \ x \in (0, 1).$$

Proof of Theorem 3.4. Let $y_0 \in H^1(0, L)$, $y_0(0) = 0$ and $y_1 \in L^2(0, L)$. Theorem 3.3 implies that there exists a Neumann control $u_N \in L^2(0, T)$, that steers the initial state

$$((y_0), \ (y_1)) \in H^1(0, L) \times L^2(0, L)$$

with the Neumann boundary condition

$$y_x^N(t, L) = u_N(t)$$

to a position of rest at the time T.

Here $y^N \in C((0, T), H^1(0, L))$ denotes the corresponding state that solves
(*NARWP*). We define

$$u(t) = u_N(t) - \gamma \, y^N(t) \in L^2(0, T).$$

Then we have

$$y_x^N(t, L) - \gamma \, y^N(t) = u(t).$$

Hence with the control u the state y^N is a solution of (*RARWP*) that satisfies the end
conditions at the time T. Thus we have shown Theorem 3.4.$\square$

Chapter 4
Optimal Exact Control

In optimal control problems, we choose the *'best'* controls from the set of all admissible controls. In our case, the set of admissible controls consists of the set of all controls that steer the system to the desired terminal state at the given terminal time. In general, these exact controls are not uniquely determined. Therefore we can choose from the set of admissible controls an exact control that is optimal in the sense that it minimizes an objective function that models our preferences. This leads to an optimal control problem where the prescribed end conditions can be regarded as equality constraints. Often, the control costs that are given by the norm of the control function are an interesting objective function.

4.1 Optimal Dirichlet control

First we consider the case of Dirichlet control. We are looking for an admissible control that has minimal L^2-norm. For sufficiently regular initial data we can also consider controls with minimal L^p-norm for $p > 2$ (see Theorem 3.1).

Let the length $L = 1$ and the wave speed $c = 1$ be given. We consider the initial state (y_0, y_1) that satisfies the regularity assumptions

$$y_0 \in L^2(0, 1), \ y_1 \in H^{-1}(0, 1).$$

Then for $T \geq 2$ there are controls $u \in L^2(0, T)$ that steer the system that is governed by the wave equation

$$y_{tt}(t, x) = y_{xx}(t, x), \ (t, x) \in (0, T) \times (0, 1)$$

and the boundary conditions

$$y(t, 0) = 0, \ y(t, 1) = u(t), \ t \in (0, T)$$

© The Author(s) 2015
M. Gugat, *Optimal Boundary Control and Boundary Stabilization of Hyperbolic Systems*, SpringerBriefs in Electrical and Computer Engineering,
DOI 10.1007/978-3-319-18890-4_4

to the position of rest exactly at the time T, that is such that

$$y(T, x) = 0, \ y_t(T, x) = 0, \ x \in (0, 1).$$

This is a consequence of Theorem 3.1 and can be seen as follows: We extend the given initial position y_0 to an even function on the interval $[-L, L]$. The initial velocity y_1 is extended to an odd function on $[-L, L]$. If the space interval is extended from $[0, L]$ to $[-L, L]$, the length of the interval is doubled. Hence also the minimal control time for control at both ends of the interval is doubled.

Equations (3.21), (3.22) for f_1 and f_2 imply: The symmetry of the initial data on $[-L, L]$ that we have generated by our construction implies that

$$f_1 = -f_2.$$

Therefore for the generated state that is defined for $x \in [-L, L]$ for $t \in [0, T]$ almost everywhere we have

$$y(t, 0) = 0.$$

Therefore we can restrict the solution to the interval $[0, L]$ and get a one-sided Dirichlet control on $[0, L]$ for the terminal time $T = 2$ that satisfies the homogeneous Dirichlet boundary condition at $x = 0$.

We consider the following optimal control problem for one-sided Dirichlet boundary control:

$$\textbf{(DEC)} \begin{cases} \min \|u\|^2_{L^2(0,T)} \text{ subject to the constraints} \\[2mm] y(0, x) = y_0(x), \ y_t(0, x) = y_1(x), \ x \in (0, 1) \\[2mm] y(t, 0) = 0, \ y(t, 1) = u(t), \ t \in (0, T) \\[2mm] y_{tt}(t, x) = y_{xx}(t, x), \ (t, x) \in (0, T) \times (0, 1) \\[2mm] y(T, x) = 0, \ y_t(T, x) = 0, \ x \in (0, 1). \end{cases}$$

In [33] the following result is presented (for the more general case $T \in (2, \infty)$):

Theorem 4.1 (Optimal Dirichlet control). *Let $T = 2K$ be an even number with $K \in \{1, 2, 3, \ldots\}$. Then the optimal control is 2 periodic.*

*Let $r = \int_0^1 \int_0^t y_1(s) \, ds \, dt$. For the solution u of **(DEC)** we have*

$$u(t) = \begin{cases} \frac{1}{T}\left(-\int_0^{1-t} y_1(s)\, ds + r + y_0(1-t)\right), \ t \in (0, 1), \\[3mm] \frac{1}{T}\left(-\int_0^{t-1} y_1(s)\, ds + r - y_0(t-1)\right), \ t \in (1, 2). \end{cases}$$

Remark 4.1. In [7], it has been stated that optimal controls with the same structure also solve the problem of time-minimal exact control with a sufficiently large upper bound for the control norm as an additional constraint.

Remark 4.2. Starting from Theorem 4.1, in [22] the penalization of the end conditions has been studied. The penalty approach consists in omitting the end conditions as equality constraints and adding instead a term in the objective function that measures the deviation from the end conditions. In front of this penalty term there is a weighting parameter (penalty parameter) $p_{penalty} > 0$. For $p_{penalty} \to \infty$ the solutions of the corresponding optimal control problems converge to the solutions of (DEC).

Remark 4.3. Theorem 4.1 shows that the solution of (DEC) is given by a linear map that maps the initial data (y_0, y_1) to the optimal control u. Since the optimal control u also depends on the terminal time T, we denote it by

$$u(\cdot, y_0, y_1, T).$$

Proof of Theorem 4.1.

Step 1: Transformation of the objective function

For $u \in L^2(0, T)$, $k \in \{1, 2, \ldots, K\}$ and $t \in (0, 2)$ define

$$u_k(t) = u(2(K - k) + t).$$

Using the u_k, the objective function of (DEC) can be represented as follows:

$$\int_0^{2K} u(t)^2 \, dt = \frac{1}{2} \int_0^1 \sum_{k=1}^{K} [u_k(1 + x) + u_k(1 - x)]^2 + [u_k(1 + x) - u_k(1 - x)]^2 \, dx.$$

Step 2: Transformation of the end conditions

Now we transform the end conditions

$$y(T, x) = 0, \ y_t(T, x) = 0, \ x \in (0, 1).$$

For this purpose we use the representation of the state y from Theorem 2.1. For $x \in (0, 1)$ almost everywhere we have

$$y(T, x) = \frac{1}{2} [\alpha(T + x) + \beta(T + 1 - x)]$$

$$= \frac{1}{2} [\alpha_{2K}(x) + \beta_{2K}(1 - x)]$$

$$= \frac{1}{2} \left[\sum_{i=0}^{(2K-2)/2} 2u(x + (2i + 1)) + \alpha_0(x) \right.$$

$$\left. + \sum_{i=0}^{(2K-2)/2} -2u((1 - x) + 2i) + \beta_0(1 - x) \right]$$

$$= \frac{1}{2} [\alpha_0(x) + \beta_0(1 - x)] + \sum_{i=0}^{K-1} [u(x + (2i + 1)) - u((1 - x) + 2i)].$$

Theorem 2.1 implies

$$\frac{1}{2} [\alpha_0(x) + \beta_0(1 - x)] = y_0(x).$$

Thus we get

$$y(T, x) = y_0(x) + \sum_{i=0}^{K-1} [u(2i + (1 + x)) - u((2i + (1 - x))]$$

$$= y_0(x) + \sum_{i=0}^{K-1} [u_{K-i}(1 + x) - u_{K-i}(1 - x)]$$

$$= y_0(x) + \sum_{k=1}^{K} [u_k(1 + x) - u_k(1 - x)].$$

Now we consider the velocity. We have

$$y_t(T, x) = \frac{1}{2} [\alpha'(T + x) + \beta'(T + 1 - x)]$$

$$= \frac{1}{2} [\alpha'_{2K}(x) + \beta'_{2K}(1 - x)]$$

$$= \frac{1}{2} \left[\sum_{i=0}^{(2K-2)/2} 2u'(x + (2i + 1)) + \alpha'_0(x) \right.$$

$$\left. + \sum_{i=0}^{(2K-2)/2} -2u'((1 - x) + 2i) + \beta'_0(1 - x) \right]$$

$$= \frac{1}{2} [\alpha'_0(x) + \beta'_0(1 - x)] + \sum_{i=0}^{K-1} [u'(x + (2i + 1)) - u'((1 - x) + 2i)]$$

Theorem 2.1 implies

$$\frac{1}{2} \left[\alpha_0'(x) + \beta_0'(1 - x) \right] = y_1(x).$$

Thus we get

$$y_t(T, x) = y_1(x) + \sum_{i=0}^{K-1} \left[u'(2i + (1 + x)) - u'((2i + (1 - x))) \right]$$

$$= y_1(x) + \sum_{i=0}^{K-1} \left[u'_{K-i}(1 + x) - u'_{K-i}(1 - x) \right]$$

$$= y_1(x) + \sum_{k=1}^{K} \left[u'_k(1 + x) - u'_k(1 - x) \right].$$

By integration this implies for $x \in (0, 1)$ almost everywhere

$$\int_0^x y_t(T, s) \, ds = \int_0^x y_1(s) \, ds + \sum_{k=1}^{K} \left[u_k(1 + x) + u_k(1 - x) \right] - r$$

with a real constant r. This implies that the end conditions are equivalent to the following pointwise restrictions for the controls:

$$\sum_{k=1}^{K} \left[u_k(1 + x) - u_k(1 - x) \right] = -y_0(x), \tag{4.1}$$

$$\sum_{k=1}^{K} \left[u_k(1 + x) + u_k(1 - x) \right] = -\int_0^x y_1(s) \, ds + r. \tag{4.2}$$

These equations are pointwise equality constraints for x in $(0, 1)$ almost everywhere.

Step 3: Reduction to two parametric auxiliary problems

In order to proceed we need the following lemma:

Lemma 4.1 (see Lemma 2.7, [33]).   *Let $p \geq 2$, a natural number d and a real number g be given. The optimization problem*

$$H(p, d, g) : \quad \min_{(f_1, \ldots, f_d) \in \mathbb{R}^d} \sum_{j=1}^{d} |f_j|^p \text{ s.t. } \sum_{j=1}^{d} f_j = g.$$

has a unique solution with the components $f_j = g/d$ and the optimal value is $|g|^p / (d)^{p-1}$.

Exercise 4.1. Prove Lemma 4.1.

Solution of Exercise 4.1. *The necessary optimality condition implies that there exists a multiplier $\lambda \in \mathbb{R}$ such that for all $j \in \{1, \ldots, d\}$ we have*

$$p \, |f_j|^{p-1} \mathrm{sign}(f_j) = \lambda. \tag{4.3}$$

By taking the absolute value on both sides of the equation we get

$$p \, |f_j|^{p-1} = |\lambda|.$$

This implies

$$|f_j| = (|\lambda|/p)^{\frac{1}{p-1}}.$$

Hence for all $j,\, k \in \{1, \ldots, d\}$ we have

$$|f_j| = |f_k|.$$

Now we consider the signs. Equation (4.3) implies

$$\mathrm{sign}(f_j) = \mathrm{sign}(\lambda).$$

Thus all f_j have the same sign. Since the absolute values are also equal, for all $j,\, k \in \{1, \ldots, d\}$ we have

$$f_j = f_k.$$

With the equality constraint this yields

$$g = \sum_{j=1}^{d} f_j = d f_k$$

and we get $f_j = \frac{g}{d}$.

Now we apply Lemma 4.1 to two parametric auxiliary problems with parameters $x \in (0, 1)$ and $r \in \mathbb{R}$, namely

$$\begin{cases} \min \frac{1}{2} \sum_{k=1}^{K} [u_k(1+x) + u_k(1-x)]^2 \\ \text{subject to} \\ \sum_{k=1}^{K} [u_k(1+x) + u_k(1-x)] = -\int_0^x y_1(s)\, ds + r, \end{cases}$$

and

$$\begin{cases} \min \frac{1}{2} \sum_{k=1}^{K} [u_k(1+x) - u_k(1-x)]^2 \\ \text{subject to} \\ \sum_{k=1}^{K} [u_k(1+x) - u_k(1-x)] = -y_0(x). \end{cases}$$

Lemma 4.1 implies that we have the solutions

$$u_k(1 + x) + u_k(1 - x) = -\frac{1}{K}\left[\int_0^x y_1(s)\,ds - r\right], \tag{4.4}$$

$$u_k(1 + x) - u_k(1 - x) = -\frac{1}{K}\,y_0(x). \tag{4.5}$$

These equations are equivalent to

$$u_k(1 + x) = \frac{1}{2K}\left[-y_0(x) + r - \int_0^x y_1(s)\,ds\right], \tag{4.6}$$

$$u_k(1 - x) = \frac{1}{2K}\left[y_0(x) + r - \int_0^x y_1(s)\,ds\right]. \tag{4.7}$$

For a given value of r, the control defined by (4.6), (4.7) solves for x in $(0, 1)$ almost everywhere both parametric auxiliary problems, where the objective functions are given by the terms in the integrand of the transformed objective function from Step 1 that are minimized subject to the corresponding equality constraints (4.1), (4.2) from Step 2 that are equivalent to the end conditions.

Since integration is a monotone operation, the control remains optimal if the objective functions of the parametric auxiliary problems are integrated on $(0, 1)$. Thus for a given value of r, the control defined by (4.6), (4.7) also solves the problem where the integrand of the objective function of **(DEC)** is integrated over all $x \in (0, 1)$ (see Step 1), that is

$$\begin{cases} \min \frac{1}{2}\int_0^1 \sum_{k=1}^K [u_k(1+x) + u_k(1-x)]^2 + \sum_{k=1}^K [u_k(1+x) - u_k(1-x)]^2\,dx \\ \text{subject to} \\ \sum_{k=1}^K [u_k(1+x) + u_k(1-x)] = -\int_0^x y_1(s)\,ds + r, \quad (x \in (0,1)) \\ \sum_{k=1}^K [u_k(1+x) - u_k(1-x)] = -y_0(x), \quad (x \in (0,1)). \end{cases}$$

In order to solve **(DEC)**, it only remains to choose the real number r in such a way that the objective function is minimal.

Exercise 4.2. Show that with the choice of r as in Theorem 4.1, we get an admissible control with minimal L^2-norm.

Solution of Exercise 4.2. *We compute the values of the objective function as a function of r by inserting for u_k the values from (4.6), (4.7). We get the objective values*

$$v(r) = \frac{1}{4K}\int_0^1 [r - y_0(x) - \int_0^x y_1(s)\,ds]^2 + [r + y_0(x) - \int_0^x y_1(s)\,ds]^2\,dx$$

$$= \frac{1}{2K}\int_0^1 (r - \int_0^x y_1(s)\,ds)^2 + (y_0(x))^2\,dx.$$

To get the minimum of $v(r)$ we consider the equation $v'(r) = 0$. This yields

$$r = \int_0^1 \int_0^x y_1(s) \, ds \, dx$$

as in Theorem 4.1.

Thus we have proved Theorem 4.1.$\square$

4.2   Optimal Neumann control

In this section we look at the problem of optimal exact control to a position of rest in a given finite time for a system that is governed by the initial boundary value problem (*NARWP*). Let an initial position $y_0 \in H^1(0, 1)$ with $y_0(0) = 0$ and an initial velocity $y_1 \in L^2(0, 1)$ be given. Let $T \geq 2$ be given. We consider the problem of optimal exact control (with $c = 1 = L$)

(NEC) $\begin{cases} \min_{u \in L^2(0,T)} \|u\|^2_{L^2(0,T)} \text{ subject to} \\[2mm] y(0, x) = y_0(x), \; y_t(0, x) = y_1(x), \; x \in (0, 1) \\[2mm] y(t, 0) = 0, \; y_x(t, 1) = u(t), \; t \in (0, T) \\[2mm] y_{tt}(t, x) = y_{xx}(t, x), \; (t, x) \in (0, T) \times (0, 1) \\[2mm] y(T, x) = 0, \; y_t(T, x) = 0, \; x \in (0, 1). \end{cases}$

The objective function in **(NEC)** is the L^2-norm of the control that can be regarded as control cost. The objective function is equal to the L^2-norm of the normal derivative of the solution at $x = 1$. Theorem 4.2 gives an explicit representation of the solution of **(NEC)** for arbitrary $T \geq 2$.

Theorem 4.2 (Representation of the optimal Neumann control, see [26]). *Let $T \geq 2$ be given. Define $k = \max\{n \in \mathbb{N} : 2n \leq T\}$ and $\Delta = T - 2k$.*
For $t \in [0, 2)$, let

$$d(t) = \begin{cases} k + 1, \; t \in (0, \Delta], \\ k, \qquad t \in (\Delta, 2). \end{cases} \tag{4.8}$$

*Then the optimal control u_0 that solves **(NEC)** is 4-periodic, with*

$$u_0(t) = \begin{cases} \frac{1}{2d(t)} \left[y_0'(1 - t) - y_1(1 - t) \right], \; t \in (0, 1), \\[3mm] \frac{1}{2d(t)} \left[y_0'(t - 1) + y_1(t - 1) \right], \; t \in (1, 2). \end{cases}$$

For $t \in (0, 2)$, $l \in \{0, 1, \ldots, k\}$ with $t + 2l \leq T$ we have

$$u_0(t + 2l) = (-1)^l u_0(t). \tag{4.9}$$

Remark 4.4. The idea of *moving-horizon control* is to use at each moment the optimal control value $u_0(0)$ that corresponds to the present state considered as the initial state. With the optimal control u_0 from Theorem 4.2 that solves **(NEC)** this yields

$$y_x(t, 1) = \frac{1}{2d(0) - 1} (-y_t(t, 1)).$$

This is a well-known exponentially stabilizing feedback law (see [38]).

Remark 4.5. The optimal value of **(NEC)** is

$$\int_0^T (u_0(t))^2 \, dt = \frac{1}{4} \int_0^1 \left[\frac{1}{d(1 - s)} + \frac{1}{d(1 + s)} \right] \left[[y_0'(s)]^2 + [y_1(s)]^2 \right] ds$$

$$- \frac{1}{2} \int_0^1 \left[\frac{1}{d(1 - s)} - \frac{1}{d(1 + s)} \right] y_0'(s) \, y_1(s) \, ds$$

$$=: \omega(T).$$

If $\Delta = 0$, that is for $T = 2k$ we get

$$\omega(2k) = \frac{1}{2k} \int_0^1 [y_0'(s)]^2 + [y_1(s)]^2 \, ds.$$

Remark 4.6. Also problems of time-minimal Neumann control for the wave equation with a sufficiently large upper bound for the control norm as control constraint have been studied. For example in [3, 51] problems with upper bounds on the L^∞ norms of Neumann controls at both ends of the space interval have been considered. It turns out that for a rest to rest maneuver with Neumann control action at both boundary points of the space interval that steers the system from a constant state with zero velocity to another constant state with zero velocity, the time optimal controls have a bang-off-bang structure, that is the time optimal controls attain only three values. One of the values (the off-value) is zero. For more examples of L^∞-norm minimal Neumann control see [18] (Figure 4.1).

Proof of Theorem 4.2 for the case $T = 2K$. For the case $T = 2K$ where K is a natural number we present a proof that is based upon the traveling waves representation (3.44) of the solution of (*NARWP*) from Theorem 3.2. (A proof for the general case is given below.)

First we have to verify that u_0 is an admissible control. Equation (3.46) implies for the position at the terminal time T for $x \in (0, 1)$ almost everywhere

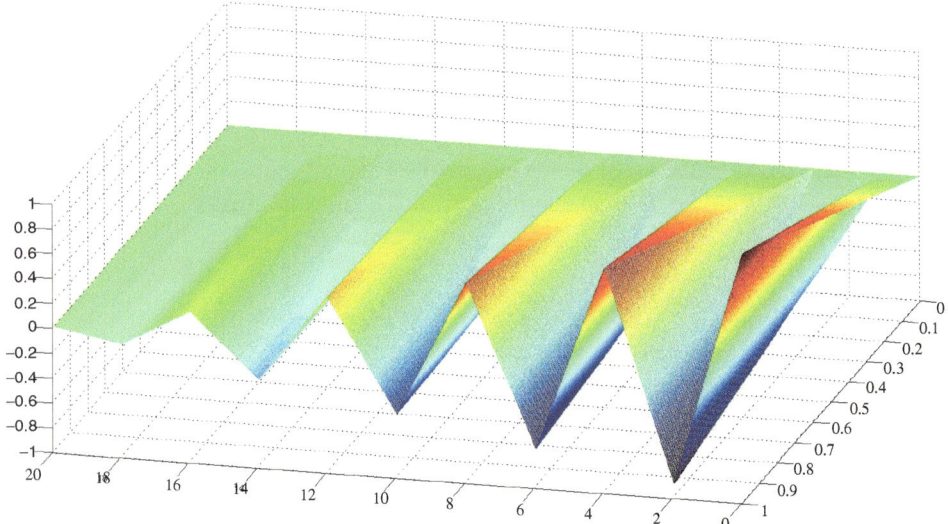

Fig. 4.1 The state y generated by the optimal Neumann boundary control that solves **(NEC)** for $L = 1$, $(y_0, y_1) = (x, 0)$ and $T = 20$.

$$y(T, x) = (-1)^K y_0(x) - \sum_{k=1}^{K} (-1)^k \int_{1-x}^{1+x} u_0(2(K - k) + s)\, ds$$

$$= (-1)^K y_0(x) - \sum_{k=1}^{K} (-1)^k (-1)^{K-k} \int_{1-x}^{1+x} u_0(s)\, ds$$

$$= (-1)^K y_0(x) - (-1)^K K \int_{1-x}^{1+x} u_0(s)\, ds$$

$$= (-1)^K y_0(x) - (-1)^K K \left[\frac{y_0(x)}{K} \right]$$

$$= 0.$$

For the velocity at the terminal time T we get on account of the sign changes of u_0 according to (4.9) and with (3.49)–(3.50) for $x \in (0, 1)$ almost everywhere

$$y_t(T, x) = (-1)^K y_1(x) - \sum_{k=1}^{K} (-1)^k (-1)^{K-k} [u_0(1 + x) - u_0(1 - x)]$$

$$= (-1)^K \left[y_1(x) - K \left(u_0(1 + x) - u_0(1 - x) \right) \right]$$

$$= (-1)^K \left[y_1(x) - K \left(\frac{y_1(x)}{K} \right) \right]$$

$$= 0.$$

Hence the control u_0 from Theorem 4.2 is admissible for **(NEC)** in the sense that it generates a state that satisfies the prescribed terminal constraints. It remains to show that u_0 is also optimal, which is the element of the set of admissible controls with minimal L^2-norm. Here the arguments are similar as in the proof of Theorem 4.1. We use the following lemma that is similar to Lemma 4.1.

Lemma 4.2 (see [33]). *Let $p \geq 2$, a natural number d and a real number q be given. The optimization problem*

$$H(p,d,g): \quad \min_{(f_1,\ldots,f_d)\in\mathbb{R}^d} \sum_{j=1}^{d} |f_j|^p \quad \text{s.t.} \quad \sum_{j=1}^{d}(-1)^j f_j = g$$

has a unique solution with the components $f_j = (-1)^j g/d$ and the optimal value is $|g|^p/(d)^{p-1}$.

Exercise 4.3. Prove Lemma 4.2.

For a control $u \in L^2(0, 2K)$, $t \in (0, 2)$ and $k \in \{1, 2, \ldots, K\}$ we define

$$u_k(t) = u(2(K - k) + t).$$

For $x \in (0, 1)$ almost everywhere Lemma 4.2 yields the solutions of the two parametric optimization problems with parameter x

$$\begin{cases} \min \frac{1}{2} \sum_{k=1}^{K} [u_k(1+x) + u_k(1-x)]^2 \\ \text{subject to} \\ \sum_{k=1}^{K}(-1)^k [u_k(1+x) + u_k(1-x)] = (-1)^K \partial_x y_0(x) \end{cases}$$

and

$$\begin{cases} \min \frac{1}{2} \sum_{k=1}^{K} [u_k(1+x) - u_k(1-x)]^2 \\ \text{subject to} \\ \sum_{k=1}^{K}(-1)^k [u_k(1+x) - u_k(1-x)] = (-1)^K y_1(x). \end{cases}$$

The definition of u_0 implies that u_0 solves both optimization problems.

This implies that for $x \in (0, 1)$ almost everywhere u_0 also solves the optimization problem where we add both objective functions and require both constraints, that is

$$\begin{cases} \min \frac{1}{2} \sum_{k=1}^{K} [u_k(1+x) + u_k(1-x)]^2 + [u_k(1+x) - u_k(1-x)]^2 \\ \text{subject to} \\ \sum_{k=1}^{K}(-1)^k [u_k(1+x) + u_k(1-x)] = (-1)^K \partial_x y_0(x) \\ \text{und} \quad \sum_{k=1}^{K}(-1)^k [u_k(1+x) - u_k(1-x)] = (-1)^K y_1(x). \end{cases}$$

Up to now, x is a parameter in the optimization problems, and the solution $u_0(x)$ as a function of x has L^2 regularity.

Now we integrate the objective function over $x \in (0, 1)$ and require that both constraints hold almost everywhere in the interval $(0, 1)$:

$$
\begin{cases}
\min_{u \in L^2(0, 2K)} \dfrac{1}{2} \int_0^1 \sum_{k=1}^K [u_k(1 + x) + u_k(1 - x)]^2 + [u_k(1 + x) - u_k(1 - x)]^2 \, dx \\[2mm]
\text{subject to} \\
\sum_{k=1}^K (-1)^k [u_k(1 + z) + u_k(1 - z)] = (-1)^K \partial_x y_0(z) \\
\text{and} \ \ \sum_{k=1}^K (-1)^k [u_k(1 + z) - u_k(1 - z)] = (-1)^K y_1(z) \\
\text{for} \ z \in (0, 1) \ \text{almost everywhere.}
\end{cases}
$$

$$(4.10)$$

Then the monotonicity property of the integral implies that u_0 also solves this optimization problem. For the objective function due to the definition of the u_k we have

$$
\frac{1}{2} \int_0^1 \sum_{k=1}^K [u_k(1 + x) + u_k(1 - x)]^2 + [u_k(1 + x) - u_k(1 - x)]^2 \, dx = \int_0^{2K} u(t)^2 \, dt.
$$

On account of (3.47)–(3.48) and (3.49)–(3.50) the terminal constraints

$$
y(T, x) = 0, \ y_t(T, x) = 0, \ x \in (0, 1)
$$

are equivalent to the constraints of the optimization problem (4.10). This implies that u_0 is the optimal control and Theorem 4.2 is proved for the case $T = 2K$.

Proof for the general case. For the general case $T > 2$ we give a proof that is based upon the method of moments. The proof consists of three steps.

Let $u^* \in L^2(0, T)$ be a control that is admissible for **(NEC)**, that is the generated state satisfies the desired end conditions at the terminal time T.

Step 1: For $t \in (0, 2)$ we define

$$
\tilde{u}(t) = \frac{1}{d(t)} \sum_{j=0}^{d(t)-1} (-1)^j u^*(t + 2j)
$$

and for $l \in \{0, 1, \ldots, d(t) - 1\}$ we define

$$
\tilde{u}(t + 2l) = (-1)^l \tilde{u}(l). \tag{4.11}
$$

The definition implies $\tilde{u} \in L^2(0, T)$ and on account of the convexity of the L^2-norm we get the inequality

$$
\|\tilde{u}\|_{L^2(0,T)}^2 \le \|u^*\|_{L^2(0,T)}^2. \tag{4.12}
$$

This can be seen as follows. Due to the convexity of the real function $u \mapsto u^2$ we get

$$\|\tilde{u}\|^2_{L^2(0,T)} = \int_0^T (\tilde{u}(s))^2 \, ds$$

$$= \sum_{j=0}^k \int_0^\Delta (\tilde{u}(s))^2 \, ds + \sum_{j=0}^{k-1} \int_\Delta^2 (\tilde{u}(s))^2 \, ds$$

$$= \int_0^2 \sum_{j=0}^{d(s)-1} (\tilde{u}(s))^2 \, ds$$

$$= \int_0^2 d(s)(\tilde{u}(s))^2 \, ds$$

$$= \int_0^2 d(s) \left(\sum_{j=0}^{d(s)-1} \frac{1}{d(s)} (-1)^j u^*(s+2j) \right)^2 ds$$

$$\leq \int_0^2 d(s) \sum_{j=0}^{d(s)-1} \frac{1}{d(s)} \left(u^*(s+2j) \right)^2 ds$$

$$= \int_0^T \left(u^*(s) \right)^2 \, ds.$$

Remark 4.7. If u^* satisfies (4.11), we have $u^* = \tilde{u}$.

Step 2: Now we show that the fact that u^* is admissible for **(NEC)** implies that $\tilde{u}$ is admissible. We consider the solution of *(NARWP)* from Theorem 2.3. The end conditions are equivalent to the sequence of moment equations

$$\alpha_n(T) = 0, \ \alpha'_n(T) = 0, \ n \in \{1,2,3,\ldots\}.$$

The moment equations have the form

$$\int_0^T \tilde{u}(s) \sin((\frac{\pi}{2} + n\pi)(T-s)) \, ds = c_j^s(T),$$

$$\int_0^T \tilde{u}(s) \cos((\frac{\pi}{2} + n\pi)(T-s)) \, ds = c_j^c(T), \ j \in \{1,2,3,\ldots\}.$$

For all $j \in \{1,2,3,\ldots\}$ we have

$$\int_0^T \tilde{u}(s) \sin((\frac{\pi}{2} + n\pi)(T-s)) \, ds = \int_0^T u^*(s) \sin((\frac{\pi}{2} + n\pi)(T-s)) \, ds,$$

$$\int_0^T \tilde{u}(s) \cos((\frac{\pi}{2} + n\pi)(T-s)) \, ds = \int_0^T u^*(s) \cos((\frac{\pi}{2} + n\pi)(T-s)) \, ds.$$

For the sin-case this can be seen as follows:

$$\int_0^T u^*(s) \sin((\frac{\pi}{2} + n\pi)(T-s))\,ds$$

$$= \sum_{j=0}^k \int_0^\Delta u^*(s+2j) \sin((\frac{\pi}{2} + n\pi)(T-s-2j))\,ds$$

$$+ \sum_{j=0}^{k-1} \int_\Delta^2 u^*(s+2j) \sin((\frac{\pi}{2} + n\pi)(T-s-2j))\,ds$$

$$= \int_0^2 \sum_{j=0}^{d(s)-1} u^*(s+2j)\ \sin((\frac{\pi}{2} + n\pi)(T-s-2j))\,ds$$

$$= \int_0^2 \sum_{j=0}^{d(s)-1} u^*(s+2j)(-1)^j\ \sin((\frac{\pi}{2} + n\pi)(T-s))\,ds$$

$$= \int_0^2 d(s)\,\tilde{u}(s)\ \sin((\frac{\pi}{2} + n\pi)(T-s))\,ds.$$

Moreover (4.11) implies

$$\int_0^2 d(s)\,\tilde{u}(s)\ \sin((\frac{\pi}{2} + n\pi)(T-s))\,ds$$

$$= \int_0^2 \sum_{j=0}^{d(s)-1} \tilde{u}(s+2j)(-1)^j\ \sin((\frac{\pi}{2} + n\pi)(T-s))\,ds$$

$$= \int_0^T \tilde{u}(s) \sin((\frac{\pi}{2} + n\pi)(T-s))\,ds.$$

For the cos-case the equations are similar. Hence the admissibility of u^* is equivalent to the admissibility of $\tilde{u}$. Step 1 implies: There is a solution of (NEC) that satisfies (4.11).

Step 3: Theorem 3.2 implies that for $x \in (0, 1)$ almost everywhere we have

$$y_t(T, x) = \alpha'(2k + \Delta + x) - \alpha'(2k + \Delta - x),$$
$$y_x(T, x) = \alpha'(2k + \Delta + x) + \alpha'(2k + \Delta - x).$$

Hence the end conditions hold if for $x \in (0, 1)$ almost everywhere we have

$$\alpha'(2k + \Delta - x) = 0 \text{ and } \alpha'(2k + \Delta + x) = 0.$$

This is equivalent to

$$\alpha'(2k + x) = 0 \ \text{ für } x \in (\Delta - 1, \ \Delta + 1).$$

Lemma 3.1 implies that for $t \in (0, 2)$ almost everywhere we have

$$\alpha(2k + t + 1) = -\alpha(2k + t - 1) + \int_0^{2k+t} u(s) \, ds$$

and thus

$$\alpha'(2k + 1 + t) = -\alpha'(2k - 1 + t) + u(t + 2k). \tag{4.13}$$

Hence the end conditions are also equivalent to

$$\alpha'(2k + x) = 0 \ \text{ for } x \in (\Delta - 1, \ 1)$$

and

$$\alpha'(2k - 1 + t) - u(t + 2k) = 0 \ \text{ for } t \in (0, \Delta).$$

Using the indicator function of the interval $(-1, -1 + \Delta)$

$$\chi(x) = 1_{(-1, -1+\Delta)}(x)$$

the end conditions can be characterized by the following equation

$$\alpha'(2k + x) - \chi(x) \, u(1 + x + 2k) = 0 \ \text{ for } x \in (-1, \ 1). \tag{4.14}$$

Using Theorem 3.2 from the state $(y(2k, x), y_t(2k, x))$ we can determine the function $\alpha'(2k + \cdot)$. For this purpose we consider the time $2k$ as initial time and compute the corresponding function $\alpha(2k + \cdot)$ as in Lemma 3.1.

For $t \in (-1, 0)$ we have

$$\alpha(2k + t) = \frac{1}{2} \left(-y(2k, -t) - \int_{-t}^1 y_t(2k, s) \, ds \right)$$

and for $t \in (0, 1)$ we have

$$\alpha(2k + t) = \frac{1}{2} \left(y(2k, t) - \int_t^1 y_t(2k, s) \, ds \right).$$

We obtain

$$\alpha'(2k + t) = \begin{cases} \frac{1}{2} [y_x(2k, -t) - y_t(2k, -t)], & t \in (-1, 0), \\ \frac{1}{2} [y_x(2k, t) + y_t(2k, t)], & t \in (0, 1). \end{cases} \tag{4.15}$$

With (4.14) we get the equations

$$0 = \frac{1}{2} [y_x(2k, -t) - y_t(2k, -t)] - \chi(t)\, u(1 + t + 2k)\,, t \in (-1, 0) \quad (4.16)$$

$$0 = \frac{1}{2} [y_x(2k, t) + y_t(2k, t)] - \chi(t)\, u(1 + t + 2k)\,, t \in (0, 1). \quad (4.17)$$

By transformation of the variables in the first equation this yields

$$0 = \frac{1}{2} [y_x(2k, t) - y_t(2k, t)] - \chi(-t)\, u(1 - t + 2k),\ t \in (0, 1) \quad (4.18)$$

$$0 = \frac{1}{2} [y_x(2k, t) + y_t(2k, t)] - \chi(t)\, u(1 + t + 2k),\ t \in (0, 1). \quad (4.19)$$

By adding the two equations and subtracting the first from the second equation we get for $t \in (0, 1)$ almost everywhere

$$0 = y_x(2k, t) - \chi(-t)\, u(2k + 1 - t) - \chi(t)\, u(2k + 1 + t), \quad (4.20)$$

$$0 = y_t(2k, t) + \chi(-t)\, u(2k + 1 - t) - \chi(t)\, u(2k + 1 + t). \quad (4.21)$$

Theorem 3.2 implies that for controls u that satisfy (4.11) we have (see Step 2)

$$y_t(2k, x) = (-1)^k y_1(x) - \sum_{j=1}^{k} (-1)^j\, (-1)^{k-j}\, [u(1 + x) - u(1 - x)]$$

$$= (-1)^k\, [y_1(x) - k\, (u(1 + x) - u(1 - x))]\,.$$

Hence (4.21) implies for $x \in (0, 1)$ almost everywhere

$$0 = y_1(x) - k\, (u(1 + x) - u(1 - x)) + \chi(-x)\, u(1 - x) - \chi(x)\, u(1 + x).$$

Thus we have

$$y_1(x) = (k + \chi(x))\, u(1 + x) - (k + \chi(-x))\, u(1 - x). \quad (4.22)$$

Now we consider y_x. Theorem 3.2 implies that for all controls u that satisfy (4.11) (see Step 2) we have

$$y_x(T, x) = (-1)^k \partial_x y_0(x) - \sum_{j=0}^{k-1} (-1)^{k-j}\, [u(2j + (1 + x)) + u(2j + (1 - x))]$$

$$= (-1)^k \partial_x y_0(x) - \sum_{j=0}^{k-1} (-1)^{k-j}(-1)^j\, [u(1 + x) + u(1 - x)]$$

$$= (-1)^k \left[\partial_x y_0(x) - \sum_{j=0}^{k-1} [u(1 + x) + u(1 - x)] \right].$$

Hence (4.20) implies that for $x \in (0, 1)$ almost everywhere we have

$$0 = \partial_x y_0(x) - k\left(u(1 + x) + u(1 - x)\right) - \chi(-x)\, u(1 - x) - \chi(x)\, u(1 + x).$$

This yields

$$\partial_x y_0(x) = (k + \chi(x))\, u(1 + x) + (k + \chi(-x))\, u(1 - x). \tag{4.23}$$

By adding (4.22) and (4.23) and by subtracting the first from the second equation we get for $t \in (0, 1)$ almost everywhere

$$u(1 + x) = \frac{1}{2(k + \chi(x))}\, [\partial_x y_0(x) + y_1(x)],$$

$$u(1 - x) = \frac{1}{2(k + \chi(-x))}\, [\partial_x y_0(x) - y_1(x)].$$

Thus we have proved Theorem 4.2. $\square$

4.3   An example for the destabilizing effect of delay

In the implementation of optimal controls, time delays can occur, for example caused by data transfer. In this section we study the effect of such delays in the implementation of optimal controls on the generated state. We consider the case that the aim of the control is to steer the energy to zero and look at the effect of the time delay on the terminal energy $E(T)$. As an example for the destabilizing effect of time delay we study optimal Dirichlet control with time delay. Let a delay $\delta > 0$ be given by a real number. We consider the control

$$u(\cdot, y_0, y_1, T)$$

that is the solution of the problem of norm-minimal exact Dirichlet control **(DEC)** and given in explicit form in Theorem 4.1. We study the situation where the control is implemented with a delay $\delta > 0$ for $t \geq \delta$ in the form

$$y(t, 1) = u(t - \delta, y_0, y_1, T). \tag{4.24}$$

For $t \in (0, \delta)$ we consider the boundary condition

$$y(t, 1) = 0. \tag{4.25}$$

Let $\delta \in (0, 1/2)$ be given. As an example for the destabilizing effect of delays in the implementation of optimal controls let us look at the initial state where

$$y_0(x) = \sin\left(\frac{\pi}{2\delta} x\right), \quad x \in (0, 1) \tag{4.26}$$

and $y_1(x) = 0$ (see [28]). Then we have $y_0(0) = 0$ and $y_0 \in H^1(0, 1)$.

The Dirichlet energy of the initial state as defined in (2.6) is

$$E(0) = \frac{1}{2} \int_0^1 y(0,x)^2 + \left(\int_0^x y_t(0,x)\,ds - \int_0^1 \int_0^z y_t(0,s)\,ds\,dz \right)^2 dx$$

$$= \frac{1}{2} \int_0^1 y_0(x)^2\,dx$$

$$= \frac{1}{2} \int_0^1 \sin^2\left(\frac{\pi}{2\delta} x\right) dx$$

$$= \frac{1}{4}\left(1 - \frac{\delta}{\pi} \sin\left(\frac{\pi}{\delta}\right)\right).$$

Lemma 4.3. *Let $\delta \in (0,1/2)$, $y_1 = 0$ and y_0 as in (4.26) be given. Let $T = 2K$ with $K \in \{1,2,3,\ldots\}$ and y be the state that is generated as the solution of (DARWP) with $c = 1 = L$ and the boundary condition (4.24), (4.25) at $x = 1$. Then for the energy we have the inequality*

$$\frac{E(T)}{E(0)} \geq \frac{1 - 2\delta + \frac{\delta}{\pi} \sin\left(\frac{\pi}{\delta}\right)}{1 - \frac{\delta}{\pi} \sin\left(\frac{\pi}{\delta}\right)}. \tag{4.27}$$

Moreover, we have

$$\lim_{\delta \to 0+} \frac{E(T)}{E(0)} \geq 2.$$

Remark 4.8. The example from Lemma 4.3 illustrates that for arbitrarily small delays $\delta > 0$ there exist initial states (y_0, y_1) with $y_1 = 0$ and $y_0 \in L^2(0,1)$ such that at the terminal time $T = 2K$ for all $K \in \{1,2,3,\ldots\}$ the energy, that is generated by the optimal control with the delay $\delta > 0$ is larger than the initial energy. More precisely, inequality (4.27) implies that for sufficiently small $\delta > 0$ we have

$$E(T) > E(0),$$

that is the energy is increased by the control. Moreover, the energy is almost doubled by the control in the sense that for all $\lambda \in (1,2)$ for $\delta > 0$ sufficiently small we have the inequality

$$E(T) \geq \lambda\, E(0).$$

Exercise 4.4. Compute an interval $I \subset (0,\infty)$, such that in the situation of Remark 4.8, for all $\delta \in I$ we have

$$E(T) > E(0).$$

Proof of Lemma 4.3. We compute the energy $E(T)$ using Theorem 2.1. Theorem 2.1 implies that for $x \in (\delta, 1 - \delta)$ almost everywhere we have

$$y(T, x) = \frac{1}{2} [\alpha(T + x) + \beta(T + (1 - x))]$$

$$= \frac{1}{2} [\alpha_{2K}(x) + \beta_{2K}(1 - x)]$$

$$= \frac{1}{2} \left[\sum_{i=0}^{(2K-2)/2} 2u(x + (2i + 1) - \delta) + \alpha_0(x) \right.$$

$$\left. + \sum_{i=0}^{(2K-2)/2} -2u((1 - x) + 2i - \delta) + \beta_0(1 - x) \right]$$

$$= \frac{1}{2} [\alpha_0(x) + \beta_0(1 - x)] + K [u(x + 1 - \delta) - u(1 - x - \delta)].$$

Theorem 2.1 implies that

$$\alpha_0(s) = y_0(s),$$
$$\beta_0(s) = y_0(1 - s)$$

and thus we have

$$\frac{1}{2} [\alpha_0(x) + \beta_0(1 - x)] = y_0(x) = \sin \left(\frac{\pi}{2\delta} x \right).$$

Theorem 4.1 states that for the optimal control we have

$$u(t) = \begin{cases} \frac{1}{T} (y_0(1 - t)), & t \in (0, 1) \\ \frac{1}{T} (-y_0(t - 1)), & t \in (1, 2) \end{cases}$$

and thus

$$u(1 + x - \delta) = \frac{1}{2K} (-y_0(x - \delta))$$

$$u(1 - x - \delta) = \frac{1}{2K} (y_0(x + \delta)).$$

This yields

$$y(T, x) = y_0(x) + \frac{1}{2} [-y_0(x + \delta) - y_0(x - \delta)]$$

$$= \sin \left(\frac{\pi}{2\delta} x \right) - \frac{1}{2} \sin \left(\frac{\pi}{2\delta} (x + \delta) \right) - \frac{1}{2} \sin \left(\frac{\pi}{2\delta} (x - \delta) \right)$$

$$= \sin \left(\frac{\pi}{2\delta} x \right) - \frac{1}{2} \sin \left(\frac{\pi}{2\delta} x + \frac{\pi}{2} \right) - \frac{1}{2} \sin \left(\frac{\pi}{2\delta} x - \frac{\pi}{2} \right)$$

$$= \sin \left(\frac{\pi}{2\delta} x \right).$$

Hence we have

$$E(T) \geq \frac{1}{2} \int_{\delta}^{1-\delta} (y(T,x))^2 \, dx = \frac{1}{2} \int_{\delta}^{1-\delta} \sin^2\left(\frac{\pi}{2\delta}x\right) dx = \frac{1}{4} - \frac{1}{2}\delta + \frac{\delta}{4\pi} \sin\left(\frac{\pi}{\delta}\right).$$

In particular, this implies

$$\liminf_{\delta \to 0+} E(T) \geq \lim_{\delta \to 0+} \frac{1}{4} - \frac{1}{2}\delta + \frac{\delta}{4\pi} \sin\left(\frac{\pi}{\delta}\right) = \frac{1}{4}.$$

Thus we have

$$\frac{E(T)}{E(0)} \geq \frac{\frac{1}{4} - \frac{1}{2}\delta + \frac{\delta}{4\pi} \sin\left(\frac{\pi}{\delta}\right)}{\frac{1}{4}\left(1 - \frac{\delta}{\pi} \sin\left(\frac{\pi}{\delta}\right)\right)}$$

$$= \frac{1 - 2\delta + \frac{\delta}{\pi} \sin\left(\frac{\pi}{\delta}\right)}{1 - \frac{\delta}{\pi} \sin\left(\frac{\pi}{\delta}\right)}.$$

This implies the inequality (4.27). Since

$$\int_0^x y_t^\delta(T,s) \, ds = \int_0^x y_1^\delta(s) \, ds + \sum_{k=1}^K [u_k(1+x-\delta) + u_k(1-x-\delta)] - r$$

we have

$$\int_0^x y_t^\delta(T,s) \, ds = K[u(1+x-\delta) + u(1-x-\delta)]$$

$$= \frac{1}{2}[y_0(x+\delta) - y_0(x-\delta)]$$

$$= \frac{1}{2}\left[\sin\left(\frac{\pi}{2\delta}x + \frac{\pi}{2}\right) - \sin\left(\frac{\pi}{2\delta}x - \frac{\pi}{2}\right)\right]$$

$$= \cos\left(\frac{\pi}{2\delta}x\right).$$

This implies (see Exercise 4.5)

$$\liminf_{\delta \to 0+} \frac{1}{2} \int_0^1 \left(\int_0^x y_t^\delta(T,x) \, ds - \int_0^1 \int_0^z y_t^\delta(T,s) \, ds \, dz\right)^2 dx$$

$$\geq \lim_{\delta \to 0+} \frac{1}{2} \int_\delta^{1-\delta} \left(\int_0^x y_t^\delta(T,x) \, ds - \frac{1}{1-2\delta} \int_\delta^{1-\delta} \int_0^z y_t^\delta(T,s) \, ds \, dz\right)^2 dx$$

$$= \lim_{\delta \to 0+} \frac{1}{2} \int_\delta^{1-\delta} \left(\cos\left(\frac{\pi}{2\delta}x\right) - \frac{1}{1-2\delta}\frac{2\delta}{\pi}\left[-\cos\left(\frac{\pi}{2\delta}\right) - 1\right]\right)^2 dx$$

$$= \lim_{\delta \to 0+} \frac{1}{2} \int_{\delta}^{1-\delta} \left(\cos\left(\frac{\pi}{2\delta}x\right) + \frac{1}{1-2\delta}\frac{2\delta}{\pi}\left[1 + \cos\left(\frac{\pi}{2\delta}\right)\right] \right)^2 dx$$

$$= \lim_{\delta \to 0+} \frac{1}{2} \int_{0}^{1} \cos^2\left(\frac{\pi}{2\delta}x\right) dx$$

$$= \lim_{\delta \to 0+} \frac{1}{4} + \frac{1}{4}\frac{\delta}{\pi} \sin\left(\frac{\pi}{\delta}\right)$$

$$= \frac{1}{4}.$$

Hence we have

$$\liminf_{\delta \to 0+} E(T) \geq \lim_{\delta \to 0+} \frac{1}{4} - \frac{1}{2}\delta + \frac{\delta}{4\pi} \sin\left(\frac{\pi}{\delta}\right) + \frac{1}{4} = \frac{1}{4} + \frac{1}{4} = \frac{1}{2}.$$

Thus we have proved Lemma 4.3. □

Exercise 4.5. Show that for every function $f \in L^2(0, 1)$ and $\delta \in (0, \frac{1}{2})$ we have

$$\int_{0}^{1} \left[f(x) - \int_{0}^{1} f(s)\, ds \right]^2 dx$$

$$= \inf_{r \in \mathbb{R}} \int_{0}^{1} [f(x) - r]^2\, dx$$

$$\geq \inf_{r \in \mathbb{R}} \int_{\delta}^{1-\delta} [f(x) - r]^2\, dx$$

$$= \int_{\delta}^{1-\delta} \left[f(x) - \frac{1}{1-2\delta} \int_{\delta}^{1-\delta} f(s)\, ds \right]^2 dx.$$

Chapter 5
Boundary Stabilization

The aim of controls for boundary stabilization is to influence the system state from the points where the control action takes place in such a way that the states approaches a given desired state. Moreover, this should happen quite fast, if possible with an exponential rate. Often this is done using feedback laws, where the current observation and possibly information from the past is used to determine the control action. Often the feedback laws do not need the complete information about the current state but only partial information that is easier to observe. The observation is taken from sensors that only get information for the state at the point where they are located. In contrast to this approach, the optimal controls for a whole time interval $[0, T]$ that we have considered in Chapter 4 are based upon the complete information about a given initial state.

Often the aim of stabilization is to bring the system close to a position of rest. Some systems posses an internal damping. In this case they have a self-stabilizing property, so control action at the boundary is not necessary. In the next section we give an example for this situation. We consider a wave equation with velocity damping on the space interval $[0, L]$. An overview on the stabilization of waves on 1-d networks is given in [59].

5.1 Telegraph equation

As an example of a self-stabilizing system we consider the telegraph equation. It is a model for the flow of electrical current and voltage on an electrical transmission line.

Let $y_0 \in H^1(0, L)$ and $y_1 \in L^2(0, L)$ be given. Let $\gamma \in \mathbb{R}$. We consider the initial boundary value problem

© The Author(s) 2015   69
M. Gugat, *Optimal Boundary Control and Boundary Stabilization of Hyperbolic Systems*, SpringerBriefs in Electrical and Computer Engineering,
DOI 10.1007/978-3-319-18890-4_5

$$
(TARWP) \begin{cases}
y(0, x) & = y_0(x), & x \in (0, L) \\
y_t(0, x) & = y_1(x), & x \in (0, L) \\
y_{tt}(t, x) + \gamma \, y_t(t, x) = c^2 \, y_{xx}(t, x), & (t, x) \in (0, \infty) \times (0, L) \\
y(t, 0) & = 0, & t \in (0, \infty) \\
y_x(t, L) & = 0, & t \in (0, \infty).
\end{cases}
$$

First we consider the question of the existence of a solution of the initial boundary value problem. As in Theorem 2.3 we can consider a series representation of the solution.

Theorem 5.1. *Let* $y_0 \in H^1(0, L)$ *with* $y_0(0) = 0$, $y_1 \in L^2(0, L)$, $\gamma \in (-\pi \, c/L, \, \pi \, c/L)$.

For $n \in \{0, 1, 2, \ldots\}$ *let*

$$
\varphi_n(x) = \frac{\sqrt{2}}{\sqrt{L}} \sin\left(\left(\frac{\pi}{2} + n\pi\right)\frac{x}{L}\right),
$$

$$
\alpha_n^0 = \int_0^L y_0(x) \, \varphi_n(x) \, dx,
$$

$$
\alpha_n^1 = \int_0^L y_1(x) \, \varphi_n(x) \, dx,
$$

$$
\lambda_n = \frac{1}{L^2}\left(\frac{\pi}{2} + n\pi\right)^2,
$$

$$
\alpha_n(t) = \exp\left(-\frac{\gamma}{2}t\right)\left[\alpha_n^0 \cos\left(\sqrt{c^2\lambda_n - \frac{\gamma^2}{4}}\,t\right)\right.
$$

$$
\left. + \alpha_n^1 \frac{1}{\sqrt{c^2\lambda_n - \frac{\gamma^2}{4}}} \sin\left(\sqrt{c^2\lambda_n - \frac{\gamma^2}{4}}\,t\right)\right].
$$

Then the function

$$
y(t, x) = \sum_{n=0}^{\infty} \alpha_n(t) \, \varphi_n(x)
$$

solves the initial boundary value problem (TARWP). For all $t \in (0, T)$ *we have*

$$
y(t, \cdot) \in L^2(0, L)
$$

and

$$
\int_0^L y(t, x)^2 \, dx = \sum_{n=0}^{\infty} (\alpha_n(t))^2.
$$

For the regularity of y we have

$$y \in C([0, T], H^1(0, L)) \cap C^1([0, T], L^2(0, L)).$$

The energy satisfies the equation

$$E(t) = \frac{1}{2} \int_0^L y_x(t, x)^2 + \frac{1}{c^2} (y_t(t, x))^2 \, dx$$

$$= \frac{1}{2} \left[\sum_{n=0}^{\infty} \alpha_n(t)^2 \int_0^L (\varphi_n'(x))^2 \, dx + \frac{1}{c^2} \alpha_n'(t)^2 \right].$$

In particular the energy decays exponentially fast in the sense that for $t \geq 0$ we have

$$E(t) \leq \exp(-\gamma \, t) \, H(y_0, y_1) \tag{5.1}$$

where H is a real-valued function.

Proof of Theorem 5.1. In order to get a series representation of the solution, we consider the eigenvalue problem

$$\varphi_{xx}(x) = -\lambda \, \varphi(x), \ x \in [0, L], \ \varphi(0) = 0, \ \varphi_x(L) = 0.$$

The corresponding eigenfunctions are the φ_n (for $n \in \{0, 1, 2, \ldots\}$) from Theorem 2.3 and the eigenvalues are given by

$$\lambda_n = \frac{1}{L^2} \left(\frac{\pi}{2} + n\pi \right)^2.$$

The functions $((\varphi_n(x))_{n=0}^{\infty}$ form a complete orthonormal system in the Hilbert space $L^2(0, L)$. We define

$$\omega_n = \frac{\pi}{2} + n\pi.$$

We have

$$\varphi_n(x) = \frac{\sqrt{2}}{\sqrt{L}} \sin \left(\omega_n \frac{x}{L} \right).$$

Now we want to determine a series of functions $\alpha_n(t) : [0, T] \to \mathbb{R}$ in such a way that for the solution of (*TARWP*) we have the series representation

$$y(t, x) = \sum_{n=0}^{\infty} \alpha_n(t) \, \varphi_n(x). \tag{5.2}$$

For all $n \in \{0, 1, 2, \ldots\}$ we have

$$\int_0^L y_{tt}(t, x)\, \varphi_n(x)\, dx + \gamma \int_0^L y_t(t, x)\, \varphi_n(x)\, dx \tag{5.3}$$

$$= c^2 \int_0^L y_{xx}(t, x)\, \varphi_n(x)\, dx \tag{5.4}$$

$$= -c^2 \int_0^L y_x(t, x)\, (\varphi_n)_x(x)\, dx + c^2\, y_x(t, x)\, \varphi_n(x)|_{x=0}^L \tag{5.5}$$

$$= -c^2 \int_0^L y_x(t, x)\, (\varphi_n)_x(x)\, dx \tag{5.6}$$

$$= c^2 \int_0^L y(t, x)\, (\varphi_n)_{xx}(x)\, dx - c^2\, y(t, x)\, (\varphi_n)_x(x)|_{x=0}^L \tag{5.7}$$

$$= c^2 \int_0^L y(t, x)\, (\varphi_n)_{xx}(x)\, dx \tag{5.8}$$

$$= -c^2 \lambda_n \int_0^L y(t, x)\, \varphi_n(x)\, dx. \tag{5.9}$$

This yields the sequence of differential equations

$$\alpha_n''(t) + \gamma \alpha_n'(t) = -\frac{c^2}{L^2} \omega_n^2 \alpha_n(t)$$

with the initial conditions

$$\alpha_n(0) = \int_0^L y_0(x)\, \varphi_n(x)\, dx =: \alpha_n^0, \quad \alpha_n'(0) = \int_0^L y_1(x)\, \varphi_n(x)\, dx =: \alpha_n^1.$$

Since $\gamma \in (-\pi c/L,\, \pi c/L)$ we have

$$c^2 \lambda_n - \frac{\gamma^2}{4} > 0.$$

Therefore the solutions are

$$\alpha_n(t) = \exp\left(-\frac{\gamma}{2}t\right)\left[\alpha_n^0 \cos\left(\sqrt{c^2 \lambda_n - \frac{\gamma^2}{4}}\, t\right)\right.$$

$$\left. + \alpha_n^1 \frac{1}{\sqrt{c^2 \lambda_n - \frac{\gamma^2}{4}}} \sin\left(\sqrt{c^2 \lambda_n - \frac{\gamma^2}{4}}\, t\right)\right].$$

With these functions $\alpha_n(t)$ the solution $y(t, x)$ of (TARWP) has the series representation (5.2). We have

$$\varphi_n'(x) = \frac{\sqrt{2}}{\sqrt{L}} \frac{\omega_n}{L} \cos\left(\omega_n \frac{x}{L}\right).$$

This implies that for $m \neq n$ we have

$$\int_0^L \varphi_m'(x)\, \varphi_n'(x)\, dx = \frac{2}{L} \frac{\omega_m \omega_n}{L^2} \int_0^L \cos\left(\omega_m \frac{x}{L}\right) \cos\left(\omega_n \frac{x}{L}\right) dx = 0. \qquad (5.10)$$

This yields

$$\int_0^L (y_x(t, x))^2\, dx = \sum_{n=0}^{\infty} \alpha_n(t)^2 \int_0^L \left(\varphi_n'(x)\right)^2 dx.$$

Moreover we have

$$\int_0^L (y_t(t, x))^2\, dx = \sum_{n=0}^{\infty} \alpha_n'(t)^2 \int_0^L \left(\varphi_n(x)\right)^2 dx.$$

Now the equation for $E(t)$ and thus (5.1) follow and Theorem 5.1 is proved. □

Remark 5.1. Here for $\gamma > 0$ the energy decays exponentially fast with the rate γ. With an additional damping from the boundary point $x = L$ of the form

$$y_x(t, L) = \eta\, y_t(t, L)$$

(this works with $\eta < 0$, see Section 5.2) the system can be damped much faster: The decay rate can be increased. In fact, boundary damping still works if $\gamma < 0$ has a sufficiently small absolute value $|\gamma|$ (see [25]).

Exercise 5.1. Compute all traveling waves for the telegraph equation with $y(t, 0) = 0$ for all $t > 0$.

Solution of Exercise 5.1. Let real numbers C and v with $v \neq c$ be given. Then the traveling waves have the form

$$y(t, x) = C \left[\exp\left(\frac{\gamma v}{c^2 - v^2}(v\, t + x)\right) - \exp\left(\frac{\gamma v}{c^2 - v^2}(v\, t - x)\right)\right].$$

5.2   Neumann control

In this section we consider control action with Neumann boundary conditions. As an example, we consider again the wave equation. Since the Neumann boundary conditions are stated in terms of the space derivative of the state, in this case we need more regular states. Let $y_0 \in H^1(0, L)$, $y_0(0) = 0$ and $y_1 \in L^2(0, L)$ be given. We consider the initial boundary value problem with Neumann-feedback

$$(FARWP) \begin{cases} y(0, x) = y_0(x), & x \in (0, L) \\ y_t(0, x) = y_1(x), & x \in (0, L) \\ y_{tt}(t, x) = c^2 \, y_{xx}(t, x), & (t, x) \in (0, \infty) \times (0, L) \\ y(t, 0) = 0, & t \in (0, \infty) \\ y_x(t, L) = \eta \, y_t(t, L), & t \in (0, \infty). \end{cases}$$

The number η is called the feedback parameter. It should be chosen in such a way that the energy in the system state decays as fast as possible (with an exponential decay). First we study the question of the existence of a solution of the initial boundary value problem.

Theorem 5.2 (Neumann-velocity feedback).

Let $y_0 \in H^1(0, L)$, $y_0(0) = 0$, $y_1 \in L^2(0, L)$ and a number $\eta \in \mathbb{R}$, $\eta \neq \frac{1}{c}$ be given. Define $t_0 = \frac{L}{c}$ and let

$$f(x) = \int_0^x y_1(\sigma) \, d\sigma \in H^1(0, L)$$

denote the anti-derivative of y_1 with $f(0) = 0$. For $s \in (0, t_0)$ we define

$$\alpha(s) = y_0(cs) + \frac{1}{c} f(cs),$$

$$\alpha(s - t_0) = -y_0(L - cs) + \frac{1}{c} f(L - cs).$$

Since $\alpha(0-) = 0 = \alpha(0+)$ the definition of α implies $\alpha \in H^1(-t_0, t_0)$. Let $T > 0$.

Now we extend the domain of α and define a function $\alpha \in H^1(-t_0, T + t_0)$ recursively as follows. Let α be the continuous function, whose derivative satisfies the following equations for $k \in \{0, 1, 2, 3, \ldots\}$ and $t \in ((k-1)t_0, (k+1)t_0)$:

$$\alpha'(t + 2t_0) = \frac{c\eta + 1}{c\eta - 1} \alpha'(t). \tag{5.11}$$

Then for $t > 0$ and $x \in (0, L)$ the function

$$y(t, x) = \frac{1}{2} \left[\alpha \left(t + \frac{x}{c} \right) - \alpha \left(t - \frac{x}{c} \right) \right] \tag{5.12}$$

solves (FARWP).

Proof of Theorem 5.2. The definition of $\alpha|_{(-t_0, t_0)}$ implies that $\alpha \in H^1(-t_0, t_0)$. The recursive definition (5.11) implies that for alle $T > 0$ we have $\alpha \in H^1(-t_0, T)$. Inserting $x = 0$ in (5.12) yields

$$y(t, 0) = \frac{1}{2} [\alpha(t) - \alpha(t)] = 0.$$

Hence the boundary condition $y(t, 0) = 0$ is satisfied. Definition (5.12) implies

$$y_x(t, x) = \frac{1}{2a} \left[\alpha' \left(t + \frac{x}{c} \right) + \alpha' \left(t - \frac{x}{c} \right) \right], \tag{5.13}$$

$$y_t(t, x) = \frac{1}{2} \left[\alpha' \left(t + \frac{x}{c} \right) - \alpha' \left(t - \frac{x}{c} \right) \right]. \tag{5.14}$$

Inserting $x = L$ yields

$$y_x(t, L) - \eta y_t(t, L) = \frac{1}{2c} \left[(1 - \eta c)\alpha' \, (t + t_0) + (1 + \eta c)\alpha' \, (t - t_0) \right] = 0. \tag{5.15}$$

Hence on account of the recursion (5.11) the boundary condition at $x = L$ holds.

The function (5.12) satisfies the wave equation in the sense of distributions (see Chapter 7). More precisely the following statement holds: Define the set $\Omega = [0, \infty) \times [0, L]$. We define the family of test functions $\mathcal{T}$ as

$$\mathcal{T} = \{ \varphi \in C^2(\Omega) : \text{There is a set } Q = [t_1, t_2] \times [x_1, x_2] \subset \Omega$$

such that the support of φ is contained in the interior of $Q.\}$.

The function y satisfies the wave equation in the following weak sense:

$$\int_\Omega y_t(t, x) \, \varphi_t(t, x) \, d(t, x) = a^2 \int_\Omega y_x(t, x) \, \varphi_x(t, x) \, d(t, x) \text{ for all } \varphi \in \mathcal{T}. \tag{5.16}$$

This can be seen as follows: For all $\varphi \in \mathcal{T}$, (5.13) and (5.14) yield with integration by parts

$$\int_\Omega y_x(t, x)\varphi_x(t, x) \, d(t, x)$$

$$= \int_{x_1}^{x_2} \int_{t_1}^{t_2} \varphi_x(t, x) \frac{1}{2c} [\alpha'(t + \frac{x}{c}) + \alpha'(t - \frac{x}{c})] \, dt \, dx$$

$$= -\int_{x_1}^{x_2} \int_{t_1}^{t_2} \varphi_{xt}(t, x)\frac{1}{2c}[\alpha(t + \frac{x}{c}) + \alpha(t - \frac{x}{c})] \, dt \, dx$$

$$= -\int_{t_1}^{t_2} \int_{x_1}^{x_2} \varphi_{tx}(t, x)\frac{1}{2c}[\alpha(t + \frac{x}{c}) + \alpha(t - \frac{x}{c})] \, dx \, dt$$

$$= \int_{t_1}^{t_2} \int_{x_1}^{x_2} \varphi_t(t, x)\frac{1}{2c^2}[\alpha'(t + \frac{x}{c}) - \alpha'(t - \frac{x}{c})] \, dx \, dt$$

$$= \int_\Omega \frac{1}{c^2} \varphi_t(t, x) \, y_t(t, x) \, d(t, x).$$

This implies (5.16). For $t = 0$ we get

$$y(0, x) = \frac{1}{2} \left[\alpha \left(\frac{x}{c} \right) - \alpha \left(-\frac{x}{c} \right) \right]$$

$$= \frac{1}{2} \left[y_0(x) + \frac{1}{c} f(x) + y_0(x) - \frac{1}{c} f(x) \right] = y_0(x).$$

Moreover we have

$$y_t(0, x) = \frac{1}{2} \left[\alpha' \left(\frac{x}{c} \right) - \alpha' \left(-\frac{x}{c} \right) \right]$$

$$= \frac{1}{2} \left[c y_0'(x) + f'(x) - c y_0'(x) + f'(x) \right] = f'(x) = y_1(x).$$

Hence also the initial conditions hold. Thus we have shown that $y(t, x)$ solves (*FARWP*) and Theorem 5.2 is proved. $\square$

With the representation (5.12) of $y(t, x)$ from Theorem 5.2 and the recursion (5.11) we can answer the question whether the feedback leads to an exponential decay of the energy.

Theorem 5.3 (Energy decay). *Let* $y_0 \in H^1(0, L)$, $y_0(0) = 0$, $y_1 \in L^2(0, L)$ *and* $\eta \in \mathbb{R}$, $\eta \neq \frac{1}{c}$ *be given. Define* $t_0 = \frac{L}{c}$. *For the energy of the solution* $y(t, x)$ *of* (*FARWP*) *with* α *from Theorem 5.2 we have*

$$E(t) = \frac{1}{2} \int_0^L y_x(t, x)^2 + \frac{1}{c^2} \left(y_t(t, x) \right)^2 \, dx \tag{5.17}$$

$$= \frac{1}{4 c^2} \int_0^L \left(\alpha' \left(t + \frac{x}{c} \right) \right)^2 + \left(\alpha' \left(t - \frac{x}{c} \right) \right)^2 \, dx \tag{5.18}$$

$$= \frac{1}{4 c} \int_{t-t_0}^{t+t_0} \left(\alpha'(s) \right)^2 \, ds. \tag{5.19}$$

For $t \geq 2t_0$ *this yields*

$$E(t) = \left(\frac{c\eta + 1}{c\eta - 1} \right)^2 E(t - 2t_0). \tag{5.20}$$

If

$$\gamma = \left| \frac{c\eta + 1}{c\eta - 1} \right| < 1 \tag{5.21}$$

this implies that the energy $E(t)$ *is decreasing.*

There are constants $C_1 > 0$ *and* $\mu > 0$ *such that for all* $t > 0$ *we have*

$$E(t) \leq C_1 E(0) \exp(-\mu t), \tag{5.22}$$

that is, the energy decays exponentially.

Exercise 5.2.

a) Determine the set

$$M = \left\{ \eta \in \mathbb{R} : \left| \frac{c\eta + 1}{c\eta - 1} \right| < 1 \right\}.$$

b) What happens with the energy for $\eta = -\frac{1}{c}$?

Solution of Exercise 5.2.

a) We have $M = (-\infty, 0)$.

b) The recursion (5.20) implies that for all $t \geq 2t_0$ we have $E(t) = 0$.

Proof of Theorem 5.3. The definition (5.17) of $E(t)$ and (5.13) and (5.14) imply the representation (5.18). With substitution this yields (5.19) and the recursion (5.11) implies (5.20).

Now we show that E is decreasing Let $h > 0$ be given. Then we have

$$E(t + h) = \frac{1}{4c} \int_{t+h-t_0}^{t+h+t_0} \left(\alpha(s)' \right)^2 ds$$

$$= \frac{1}{4c} \int_{t+h-t_0}^{t+t_0} \left(\alpha(s)' \right)^2 ds + \frac{1}{4c} \int_{t-t_0}^{t+h-t_0} \left(\alpha(s + 2t_0)' \right)^2 ds$$

$$= \frac{1}{4c} \int_{t+h-t_0}^{t+t_0} \left(\alpha(s)' \right)^2 ds + \frac{1}{4c} \gamma^2 \int_{t-t_0}^{t+h-t_0} \left(\alpha(s)' \right)^2 ds$$

$$\leq \frac{1}{4c} \int_{t-t_0}^{t+t_0} \left(\alpha(s)' \right)^2 ds$$

$$= E(t).$$

Hence E is decreasing. For all $t \in [0, 2t_0)$ and for all $j \in \{0, 1, 2, \ldots\}$ we get the inequality

$$E(t + 2jt_0) \leq \gamma^j E(0). \tag{5.23}$$

For the proof of the exponential decay we need the following Lemma from [30].

Lemma 5.1 (Exponential decay). *Let $\lambda > 0$ and the function $E : [0, \infty) \to [0, \infty)$ be given. Then the following two statements are equivalent:*

1. *E decays exponentially, that is there are numbers C_1, $\mu \in (0, \infty)$ such that for all $t \in [0, \infty)$ we have:*

$$E(t) \leq C_1 E(0) \exp(-\mu t).$$

2. *There are real numbers $C_2 > 0$ and $f \in (0, 1)$ such that for all $t \in [0, \lambda)$ and for all $j \in \{0, 1, 2, \ldots\}$ the inequality*

$$E(t + j\lambda) \leq f^j C_2 E(0)$$

holds.

Proof of Lemma 5.1. First we show that 1. implies 2. Assume that 1. holds. Then for all $t \in [0, \lambda)$ and all $j \in \{0, 1, 2, \ldots\}$ we have

$$
\begin{aligned}
E(t + j\lambda) &\leq C_1 E(0) \, \exp(-\mu(t + j\lambda)) \\
&= C_1 E(0) \, \exp(-\mu t) \, \exp(-\mu \lambda j) \\
&\leq C_1 E(0) \, \exp(-\mu \lambda j) \\
&= C_1 E(0) \, \exp(-\lambda \mu)^j \\
&= f^j C_2 E(0)
\end{aligned}
$$

with $C_2 = C_1$ and $f = \exp(-\lambda \mu)$.

Now we show that 2. implies 1. Assume that 2. holds. For $j \in \{0, 1, 2, \ldots\}$ we define $t_j = j\lambda$. For all $t \in [0, \infty)$ there is a number $j \in \{0, 1, 2, \ldots\}$ such that $t \in [t_j, t_{j+1})$. Thus we have the representation

$$
t = t_j + s,
$$

with $s \in [0, \lambda)$. Define

$$
\mu = -\frac{\ln(f)}{\lambda}.
$$

Then we have $\ln(f) = -\lambda \mu$. Define $C_1 = C_2 \exp(\lambda \mu)$. Then 2. implies

$$
\begin{aligned}
E(t) &= E(s + t_j) \\
&= E(s + \lambda j) \\
&\leq f^j C_2 E(0) \\
&= \exp(j \ln(f)) \, C_2 E(0) \\
&= \exp(-j \lambda \mu) \, C_1 \exp(-\lambda \mu) \, E(0) \\
&= C_1 \exp(-\mu t_j) \, \exp(-\mu \lambda) \, E(0) \\
&\leq C_1 \exp(-\mu t_j) \, \exp(-\mu s) \, E(0) \\
&= C_1 \exp(-\mu (t_j + s)) \, E(0) \\
&= C_1 \exp(-\mu t) \, E(0)
\end{aligned}
$$

and 1. follows. Thus we have shown Lemma 5.1. $\square$

On account of (5.23), Lemma 5.1 (2. implies 1.) yields the inequality (5.22) and the proof of Theorem 5.3 is complete. $\square$

Remark 5.2. Similarly it can be shown that if for γ as defined in (5.21) we have $\gamma > 1$, the energy is increasing and blows up exponentially fast if $E(0) > 0$. For the limit case $\gamma = 1$ the energy is conserved. In this case we have $\eta = 0$ which means that the feedback control is switched off.

5.3 Time Delay

In the implementation of feedback controls time delays can occur, for example caused by the transfer of the information from the sensors to the control unit. In this section we study the possible effects of such a delay. We consider constant time delays. Again we look at a vibrating string that is fixed at one end and that is controlled by a velocity feedback at the other end. We show that for certain values of the delay time the system is exponentially stable if the feedback parameter has the right sign and sufficiently small absolute value. These results can be found in [23, 24, 55] respectively.

However, we emphasize that for many, in particular for arbitrarily small delays the system becomes unstable (see Section 5.3.4). This has been stated in the fundamental paper by Datko, Lagnese, and Polis, see [15]. In this paper and in [14] it is stated that *some second-order vibrating systems cannot tolerate small time delays in their damping.* In other words, *constant delays can destabilize a system that is asymptotically stable in the absence of delays* (see also [45]).

5.3.1 Definition of System **S**

Let the length $L > 0$, the wave speed $c > 0$, and a real number f be given. We define $\Omega = (0, \infty) \times (0, L)$ and the set of initial states

$$B = \{(y_0, y_1) \in H^1(0, L) \times L^2(0, L) : y_0(0) = 0\}.$$

For $(y_0, y_1) \in B$ we consider the system **S** :

$$y(0, x) = y_0(x), \ x \in (0, L) \tag{5.24}$$

$$y_t(0, x) = y_1(x), \ x \in (0, L) \tag{5.25}$$

$$y_{tt}(t, x) = c^2 y_{xx}(t, x), \ (t, x) \in \Omega \tag{5.26}$$

$$y(t, 0) = 0, \ t > 0 \tag{5.27}$$

$$y_x(t, L) = 0, \ t \in (0, 2\tfrac{L}{c}) \tag{5.28}$$

$$y_x(t, L) = \tfrac{f}{c} \, y_t \left(t - 2\tfrac{L}{c}, L\right), \ t > 2\tfrac{L}{c}. \tag{5.29}$$

5.3.2 Well-posedness of System **S**

In this section we consider the question of the existence of a unique solution of **S**, that is of the initial boundary value problem (5.24)–(5.29). In Theorem 5.4 we consider again traveling wave solutions.

Theorem 5.4. *Let* $(y_0, y_1) \in B$ *be given. Define the function* α *recursively by*

$$\alpha(x) = -\frac{1}{2} y_0(-x) + \frac{1}{2c} \int_0^{-x} y_1(s) \, ds, \ x \in [-L, 0), \tag{5.30}$$

$$\alpha(x) = \frac{1}{2} y_0(x) + \frac{1}{2c} \int_0^x y_1(s) \, ds, \ x \in [0, L), \tag{5.31}$$

$$\alpha(x) = \frac{1}{c} \int_0^L y_1(s) \, ds - \alpha(x - 2L), x \in [L, 3L), \tag{5.32}$$

and for $k \in \{0, 1, 2, \ldots\}$ *and* $x \in [3L + 2kL, 5L + 2kL)$ *by*

$$\alpha(x) = (f - 1)\,\alpha(x - 2L) - f\,\alpha(x - 4L) + C_k, \tag{5.33}$$

where the real constants C_k *are chosen in such a way that* α *is continuous on the interval* $[-L, \infty)$. *Let*

$$y(t, x) = \alpha(c\,t + x) - \alpha(c\,t - x), \quad (t, x) \in \Omega. \tag{5.34}$$

For every finite interval $I \subset [-L, \infty)$ *we have* $\alpha' \in L^2(I)$. *The function* y *is continuous on* Ω *and we have* $y_t, y_x \in L^1_{loc}(\Omega)$. *Define the family of test functions* $\mathcal{T}$ *as*

$$\mathcal{T} = \{\varphi \in C^2(\Omega) : \ There \ is \ a \ set \ Q = [t_1, t_2] \times [x_1, x_2] \subset \Omega$$

such that the support of φ *is contained in the interior of* $Q.\}$.

The function y *satisfies the wave equation* (5.26) *in the following weak sense: For all* $\varphi \in \mathcal{T}$ *we have*

$$\int_\Omega y_t(t, x) \varphi_t(t, x) \, d(t, x) = c^2 \int_\Omega y_x(t, x) \varphi_x(t, x) \, d(t, x). \tag{5.35}$$

The function y *satisfies* (5.24), (5.25) *and* (5.27)–(5.29). *In this sense* y *is the solution of System* **S**, *that is of the initial boundary value problem* (5.24)–(5.29).

Proof of Theorem 5.4. Since $y_0' \in L^2(0, L)$, the Sobolev-imbedding theorem implies that y_0 is continuous. Moreover we have $y_1 \in L^2(0, L)$, hence α is well defined. Now we discuss the regularity of α. On the intervals $[-L, 0)$, $[0, L)$ and $[L, 3L)$ the function α is continuous. The definition of the set B implies

$$\lim_{x \to 0-} \alpha(x) = -(1/2)y_0(0) = 0 = (1/2)y_0(0) = \lim_{x \to 0+} \alpha(x),$$

$$\lim_{x \to L-} \alpha(x) = \frac{1}{2} y_0(L) + \frac{1}{2c} \int_0^L y_1(s) \, ds$$

$$= \frac{1}{c} \int_0^L y_1(s)\, ds - \left(-\frac{1}{2} y_0(L) + \frac{1}{2c} \int_0^L y_1(s)\, ds \right)$$

$$= \frac{1}{c} \int_0^L y_1(s)\, ds - \alpha(-L) = \lim_{x \to L+} \alpha(x),$$

hence α is continuous on the interval $[-L, 3L)$. The recursion (5.33) and the choice of the constants C_k implies that the function α is continuous on the interval $[-L, \infty)$.

The derivative α' in the sense of distributions exists on the intervals $(-L, 0)$, $(0, L)$, $(L, 3L)$ and $(3L + 2kL, 5L + 2kL)$ as an L^2-function.

Since α is continuous, this implies that α is absolutely continuous on $(-L, \infty)$. Hence we have $\alpha' \in L^2_{loc}(-L, \infty)$.

The continuity of α implies the continuity of y. For $t = 0$ and $x \in (0, L)$ we have

$$y(0, x) = \alpha(x) - \alpha(-x) = y_0(x).$$

For $(t, x) \in \Omega$ almost everywhere we have

$$y_t(t, x) = c[\alpha'(x + ct) - \alpha'(-x + ct)]. \tag{5.36}$$

Hence the definition of α implies the equation $y_t(0, x) = y_1(x)$. Thus the initial conditions (5.24) and (5.25) hold.

For $(t, x) \in \Omega$ almost everywhere we have

$$y_x(t, x) = \alpha'(x + ct) + \alpha'(-x + ct). \tag{5.37}$$

With Tonelli's Theorem (see for example [46]), (5.37) implies $y_x \in L^1_{loc}(\Omega)$, and (5.36) implies $y_t \in L^1_{loc}(\Omega)$.

For all $\varphi \in \mathcal{T}$, with integration by parts (5.37) and (5.36) imply as in the Proof of Theorem 5.2

$$\int_\Omega y_x(t, x)\varphi_x(t, x)\, d(t, x)$$

$$= \int_{x_1}^{x_2} \int_{t_1}^{t_2} \varphi_x(t, x)[\alpha'(x + ct) + \alpha'(-x + ct)]\, dt\, dx$$

$$= -\int_{x_1}^{x_2} \int_{t_1}^{t_2} \varphi_{xt}(t, x)[\alpha(x + ct) + \alpha(-x + ct)]/c\, dt\, dx$$

$$= -\int_{t_1}^{t_2} \int_{x_1}^{x_2} \varphi_{tx}(t, x)[\alpha(x + ct) + \alpha(-x + ct)]/c\, dx\, dt$$

$$= \int_{t_1}^{t_2} \int_{x_1}^{x_2} \varphi_t(t, x)[\alpha'(x + ct) - \alpha'(-x + ct)]/c\, dx\, dt$$

$$= \int_\Omega \varphi_t(t, x)\, y_t(t, x)/c^2\, d(t, x).$$

Thus (5.35) holds.

For $x = 0$ we have $y(t, 0) = \alpha(ct) - \alpha(ct) = 0$, hence at $x = 0$ the boundary condition $y(t, 0) = 0$ holds for all $t \in (0, T)$.

For $x = L$, (5.37) implies for $t \in (0, 2\frac{L}{c})$

$$y_x(t, L) = \alpha'(L + ct) + \alpha'(ct - L) = -\alpha'(ct - L) + \alpha'(ct - L) = 0.$$

Hence the boundary condition (5.28) holds for all $t \in (0, 2\frac{L}{c})$. For $t > 2\frac{L}{c}$ we have

$$
\begin{aligned}
y_x(t, L) &= \alpha'(ct + L) + \alpha'(ct - L) \\
&= (f - 1)\alpha'(ct - L) - f\alpha'(ct - 3L) + \alpha'(ct - L) \\
&= f[\alpha'(L + ct - 2L) - \alpha'(-L + ct - 2L)] \\
&= (f/c)\, y_t(t - 2(L/c), L).
\end{aligned}
$$

Hence the boundary condition (5.29) holds for $t > 2\frac{L}{c}$. Thus Theorem 5.4 is proved. □

5.3.3   Exponential Stability of System S

Let us consider again the energy

$$E(t) = \frac{1}{2} \int_0^L (y_x(t, x))^2 + \frac{1}{c^2}(y_t(t, x))^2 \, dx. \tag{5.38}$$

The following result about the exponential stability of **S** is stated in [23, 24].

Theorem 5.5 (Exponential Stability of S). *For all* $f \in (0, 1)$ *System* **S** *is exponentially stable in the sense that the energy decays exponentially.*

For all $f \in (0, (\sqrt{2} - 1)^2]$ *there is a constant* $C_0 > 0$ *that depends only on* f *and on the initial state* (y_0, y_1), *such that for all natural numbers* $j \in \{0, 1, 2, \ldots\}$ *and for all* $t \in [0, 2L/c)$ *we have*

$$E(t + 2j\frac{L}{c}) \leq (1 - f)^{2j} C_0.$$

Remark 5.3. The case $f \in ((\sqrt{2} - 1)^2, 1)$ is considered in Exercise 5.3.

For the system without delay that is governed by the feedback law

$$c\, y_x(t, L) = f\, y_t(t, L), \ t > 0 \tag{5.39}$$

Theorem 5.3 implies that the energy decays exponentially for all $f < 0$.

For the feedback law (5.29) with delay the sign of the stabilizing feedback parameters f is reversed due to the additional reflection at the boundary point $x = 0$. Moreover, in the case with delay the stabilizing feedback parameters are from a bounded interval.

Proof of Theorem 5.5. Here we present the proof for $f \in (0, (\sqrt{2} - 1)^2]$.

Theorem 5.4 states that System **S** has a solution, and the corresponding energy from (5.38) has the following representation:

$$E(t) = \int_0^L \alpha'(x + ct)^2 + \alpha'(-x + ct)^2 \, dx$$

$$= \int_{-L}^L \alpha'(x + ct)^2 \, dx.$$

Let $h = 2L$. For $x \geq 3L$, (5.33) yields the equation

$$\alpha'(s) + (1 - f)\alpha'(s - h) + f\alpha'(s - 2h) = 0. \tag{5.40}$$

We consider the characteristic polynomial

$$p_f(t) = t^2 + (1 - f)t + f. \tag{5.41}$$

For $f \in (0, (\sqrt{2} - 1)^2]$ the corresponding roots are

$$z_{1,2} = -\frac{1}{2} + \frac{1}{2}f \pm \frac{1}{2}\sqrt{1 - 6f + f^2}.$$

The equation

$$1 - 6f + f^2 = (f - 3 + 2\sqrt{2})(f - 3 - 2\sqrt{2})$$

yields for all f with $f \leq 3 - 2\sqrt{2} = (1 - \sqrt{2})^2$ the inequality $1 - 6f + f^2 \geq 0$ and thus

$$|z_1| = \frac{1}{2}(1 - f - \sqrt{1 - 6f + f^2}),$$

$$|z_2| = \frac{1}{2}(1 - f + \sqrt{1 - 6f + f^2}).$$

Hence we have $|z_1| \leq |z_2| \leq 1 - f$.

With the usual method for linear difference equations we obtain an explicit representation of the solution $\alpha' \in L^2_{loc}(-L, \infty)$ of (5.40). We choose the functions $c_1(s), c_2(s)$ in such a way that for $s \in (-L, L)$ almost everywhere we have

$$\begin{pmatrix} \alpha'(s) \\ \alpha'(s + h) \end{pmatrix} = \begin{pmatrix} 1 & 1 \\ z_1 & z_2 \end{pmatrix} \begin{pmatrix} c_1(s) \\ c_2(s) \end{pmatrix}.$$

Since the matrix is invertible and since $\alpha' \in L^2_{loc}(-L, \infty)$ this implies c_1, c_2 in $L^2_{loc}(-L, L)$. Hence for all $j \in \{0, 1, 2, \ldots\}$ we have the representation

$$\alpha'(s + jh) = c_1(s)z_1^j + c_2(s)z_2^j.$$

This yields the inequality

$$|\alpha'(s + jh)| \leq (|c_1(s)| + |c_2(s)|)\,(1 - f)^j. \tag{5.42}$$

For $t \in [0, 2L/c)$ the energy satisfies the inequality

$$E(t + \frac{2jL}{c}) = \int_{-L}^{L} \alpha'(x + jh + ct)^2\,dx$$

$$= \int_{-L}^{L-ct} \alpha'(x + ct + jh)^2\,dx$$

$$+ \int_{L-ct}^{L} \alpha'(x + ct - 2L + (j+1)h)^2\,dx$$

$$= \int_{-L}^{L-ct} \left|c_1(x + ct)z_1^j + c_2(x + ct)z_2^j\right|^2\,dx$$

$$+ \int_{L-ct}^{L} \left|c_1(x + ct - 2L)z_1^{j+1} + c_2(x + ct)z_2^{j+1}\right|^2\,dx$$

$$\leq 2\,(1 - f)^{2j} \int_{-L}^{L} (|c_1(s)| + |c_2(s)|)^2\,ds.$$

Now Lemma 5.1 yields the exponential decay of the energy. Thus we have shown Theorem 5.5 for $f \in (0, (\sqrt{2} - 1)^2]$. $\square$

Exercise 5.3. Prove Theorem 5.5 for the case $f \in ((\sqrt{2} - 1)^2, 1)$: For all $f \in (\sqrt{2} - 1)^2,\ 1)$ there is a constant $C_0 > 0$ that only depends on f and the initial state (y_0, y_1) such that for all natural numbers $j \in \{0, 1, 2, \ldots\}$ and for all $t \in [0, 2L/c)$ we have

$$E(t + 2j\frac{L}{c}) \leq f^j C_0.$$

Solution of Exercise 5.3 (see [23]). As in the Proof of Theorem 5.5 we get the characteristic polynomial

$$p_f(t) = t^2 + (1 - f)\,t + f.$$

The roots are

$$z_{1,2} = \frac{1}{2}\left[(f - 1) \pm i\sqrt{6f - 1 - f^2}\right].$$

We have

$$|z_{1,2}| = \sqrt{f} < 1.$$

As in the Proof of Theorem 5.5 this yields the assertion.

Example 5.1. For $f = 1/6$ we have $z_1 = -1/3$ and $z_2 = -1/2$. For $s \in (-L, L)$ and $j \in \{0, 1, 2, \ldots\}$ this implies $\alpha'(s + jh) = (-1/3)^j c_1(s) + (-1/2)^j c_2(s)$. If y_0' and $y_1 \in L^\infty(0, 1)$, we have $\alpha' \in L^\infty(-L, \infty)$ and thus $c_1, c_2 \in L^\infty(-L, L)$. This yields the inequality

$$\|\alpha'(\cdot + jh)\|_{L^\infty(-L,L)} \leq \left(\frac{1}{2}\right)^j \left(\|c_1\|_{L^\infty(-L,L)} + \|c_2\|_{L^\infty(-L,L)}\right)$$

and Lemma 5.1 implies, that also

$$E_\infty(t) = \|\alpha'\|_{L^\infty(t-L,t+L)}$$

decays exponentially fast.

Exercise 5.4.
a) For System **S** with $f = 0$ we have $E(t) = E(0)$. Show that for System **S** with $f = 1$ for all $j \in \{0, 1, 2, \ldots\}$ and all $t \in (0, 4L/c)$ we have

$$E(t + 4jL/c) = E(t).$$

b) Show that for System **S** with $f < 0$ there exist initial states (y_0, y_1) such that there exists a constant $C_0 > 0$ that depends on (y_0, y_1) such that for all $j \in \{0, 1, 2, \ldots\}$ and all $t \in [0, 2L/c)$ we have

$$E(t + 2jL/c) \geq (1 - f)^{2j} C_0.$$

Here the delay has a destabilizing effect for negative values of the feedback parameter f!

c) Show that for System **S** with $f \in (1, \, 3 + 2\sqrt{2}]$ there exist initial states (y_0, y_1), such that there exists a constant $C_0 > 0$ that depends on (y_0, y_1) such that for all $j \in \{0, 1, 2, \ldots\}$ and all $t \in [0, 2L/c)$ we have

$$E(t + 2jL/c) \geq f^j C_0.$$

5.3.4 Destabilization of System **S** by small delays

Arbitrarily small time delays can destabilize system **S**. This can be seen as follows: Let $y(t, x)$ denote the following solution of the wave equation (with $\omega \neq 0$):

$$y(t, x) = \cos(\omega(ct + x)) - \cos(\omega(ct - x))$$
$$= -2 \sin(\omega ct) \sin(\omega x).$$

Then we have

$$y(t, 0) = \cos(\omega ct) - \cos(\omega ct) = 0.$$

Moreover, we have

$$y_x(t, x) = -2\,\omega\,\sin(\omega\,c\,t)\,\cos(\omega\,x)$$

and

$$y_t(t, x) = -2c\,\omega\,\cos(\omega\,c\,t)\,\sin(\omega\,x).$$

This implies

$$y_x(t, L) = -2\omega\sin(\omega c\,t)\,\cos(\omega\,L).$$

For the velocity we have

$$y_t(t - \delta, L) = -2\,c\,\omega\,\cos(\omega(c\,t - c\delta))\,\sin(\omega\,L).$$

Now we consider the delay

$$\delta = \frac{3\,\pi}{2\,c\,\omega}.$$

If $\cos(\omega\,L) \neq 0$ we get

$$\begin{aligned}
y_t(t - \delta, L) &= -2\,c\,\omega\,\cos(\omega(c\,t - c\delta))\,\sin(\omega\,L) \\
&= -2\,c\,\omega\,\cos(\omega\,c\,t - \frac{3\,\pi}{2})\,\sin(\omega\,L) \\
&= 2\,c\,\omega\,\sin(\omega\,c\,t)\,\sin(\omega\,L) \\
&= -c\,\tan(\omega\,L)\,y_x(t, L).
\end{aligned}$$

If we also have $\sin(\omega\,L) \neq 0$ this yields

$$y_x(t, L) = -\frac{1}{c}\,\cot(\omega\,L)\,y_t(t - \delta, L).$$

Hence $y(t, x)$ is a solution of System **S** with the feedback parameter

$$f = -\frac{1}{c}\,\cot(\omega\,L)$$

if $\sin(\omega\,L)\,\cos(\omega\,L) \neq 0$.

Since y is periodic, there cannot be an energy decay. For the energy we have

$$\begin{aligned}
E(t) &= \int_0^L \omega^2\sin^2(\omega(c\,t + x)) + \omega^2\sin^2(\omega(c\,t - x))\,dx \\
&= \omega^2 \int_0^L 1 - \cos(2\omega c t)\cos(2\omega x)\,dx
\end{aligned}$$

$$= \omega^2 \left[L - \frac{1}{2\omega} \cos(2\omega ct)\, \sin(2\omega x)|^L_{x=0} \right]$$

$$= \omega^2 \left[L - \frac{1}{2\omega} \cos(2\omega ct)\, \sin(2\omega L) \right].$$

Hence the energy does not decay to zero. For sufficiently large values of the frequency ω the delay

$$\delta = \frac{3\,\pi}{2a\,\omega}$$

becomes smaller than any given positive bound. Since the cotangent is a π-periodic function, the frequency ω can be increased by integer multiples of π/L, without changing the feedback parameter $f = -\frac{1}{c}\cot(\omega L)$. In this way we get arbitrarily small time delays that destabilize System **S**. For growing frequency ω the amplitude of the oscillation of the energy is also increased, since we have

$$E_{\max} - E_{\min} = \omega \,|\sin(2\,\omega\, L)|.$$

In contrast to this situation where the system is destabilized by arbitrarily small delays, for the system without delay that is for $\delta = 0$, Theorem 5.3 implies: For feedback parameters f that satisfy $f = -\frac{1}{c}\cot(\omega L) < 0$, the energy decays exponentially!

Chapter 6
Nonlinear Systems

Up to now, we have considered linear systems. If for such a linear system the existence of a solution can be shown for a certain finite time interval, then the solution exists for all times provided that the control keeps its regularity. For *nonlinear systems*, the situation is completely different. In a nonlinear hyperbolic system, the solution can loose a part of its regularity after a finite time. For example, classical solutions typically break down after finite time since there is a blow up in certain partial derivatives.

6.1 The Korteweg-de Vries Equation (KdV)

JOHN SCOTT RUSSELL (1808–1882), a Scottish engineer, has made the following observations about waves (*Report of the fourteenth meeting of the British Association for the Advancement of Science, York, September 1844 (London 1845), pp. 311–390, Plates XLVII-LVII*):

"I was observing the motion of a boat which was rapidly drawn along a narrow channel by a pair of horses, when the boat suddenly stopped - not so the mass of water in the channel which it had put in motion; it accumulated round the prow of the vessel in a state of violent agitation, then suddenly leaving it behind, rolled forward with great velocity, assuming the form of a large solitary elevation, a rounded, smooth and well-defined heap of water, which continued its course along the channel apparently without change of form or diminution of speed. I followed it on horseback, and overtook it still rolling on at a rate of some eight or nine miles an hour, preserving its original figure some thirty feet long and a foot to a foot and a half in height. Its height gradually diminished, and after a chase of one or two

© The Author(s) 2015
M. Gugat, *Optimal Boundary Control and Boundary Stabilization of Hyperbolic Systems*, SpringerBriefs in Electrical and Computer Engineering,
DOI 10.1007/978-3-319-18890-4_6

miles I lost it in the windings of the channel. Such, in the month of August 1834, was my first chance interview with that singular and beautiful phenomenon which I have called the Wave of Translation".

As a model for the movement of water the Korteweg-de Vries equation (KdV equation)

$$\partial_t y + y\, \partial_x y + \partial_{xxx} y = 0 \tag{6.1}$$

is considered. An overview about the research about the KdV equation is given in [48]. In *Interaction of Solitons in a Colisionless Plasma and the Recurrence of Initial States* (Phy. Rev. Let. 15, 240–243, 1965) N.J. Zabusky and M.D. Kruskal describe certain traveling waves that solve the KdV equation, so-called **solitons**.

In this chapter we consider the KdV equation on a finite space interval $[0, L]$ with boundary control action. We consider the partial differential equation

$$\partial_t y + \partial_x y + \partial_{xxx} y + y\, \partial_x y = 0 \tag{6.2}$$

with the extra term $\partial_x y$ (see [6]). With the extra term the waves move in the positive direction. Equation (6.2) can be considered as a perturbed transport equation

$$\partial_t y + \partial_x y = 0.$$

The boundary conditions are

$$y(t, 0) = u_1(t),\ y(t, L) = u_2(t),\ y_x(t, L) = u_3(t).$$

The initial condition has the form

$$y(0, x) = y_0(x),\ x \in (0, L)$$

with

$$y_0 \in L^2(0, L).$$

In the sequel we consider the boundary control with $u_1 = u_2 = 0$ and $u(t) = u_3(t) \in L^2(0, T)$ (see [12]). We consider the initial boundary value problem for small initial data, that is with an assumption of the form

$$\|y_0\|_{L^2(0,L)} \le \delta$$

with a number $\delta > 0$ that is chosen sufficiently small.

6.1.1   Well-posedness of the linearized system

We start with a result about the well-posedness of the initial value problem for the linearized system. For this purpose we consider the linearized partial differential equation

$$\partial_t y + \partial_x y + \partial_{xxx} y = \tilde{h} \tag{6.3}$$

where

$$\tilde{h} \in L^1((0, T), L^2(0, L)).$$

We consider the spatial differential operator

$$A : y \mapsto -y_x - y_{xxx} \tag{6.4}$$

with the domain

$$D(A) = \{y \in H^3(0, L) : y(0) = y(L) = y'(L) = 0\}.$$

For all $w \in D(A)$ we have

$$\int_0^L w A w \, dx = \int_0^L w(-w' - w''')$$

$$= \int_0^L (-\frac{1}{2} w^2)' + \int_0^L w' w''$$

$$= 0 + \int_0^L (\frac{1}{2} (w')^2)'$$

$$= 0 + 0 - \frac{1}{2}(w'(0))^2$$

$$\le 0.$$

If such an inequality $\int_0^L w A w \, dx \le 0$ holds whenever $w \in D(A)$, the operator A is called *dissipative* (see [9, 58]). The adjoint operator A^* is $A^* : w \mapsto w' + w'''$ with the domain

$$D(A^*) = \{y \in H^3(0, L) : y(0) = y(L) = y'(0) = 0\}.$$

This is shown using integration by parts. For all $w \in D(A^*)$ we have

$$\int_0^L w A^* w \, dx = \int_0^L w(w' + w''')$$

$$= \int_0^L (\frac{1}{2} w^2)' - \int_0^L w' w''$$

$$= 0 + 0 - \frac{1}{2}(w'(L))^2 \le 0.$$

Hence A^* is also dissipative. This implies that A generates what is called a strongly continuous semigroup of contractions (see for example Chapter 3 in [54], [4], [8]). Here is the corresponding definition:

Definition 6.1. Let X be a Banach space and $L(X)$ denote the set of linear operators on X. A strongly continuous semigroup of contractions is a map

$$\mathbb{T} : [0, \infty) \to L(X),$$

that we denote as a family of linear operators $\mathbb{T}_t$ that satisfies the following conditions:

1. $\mathbb{T}_0 = I$, that is $\mathbb{T}_0$ is the identity.
2. For all $t_1, t_2 \in [0, \infty)$ we have $\mathbb{T}_{t_1 + t_2} = \mathbb{T}_{t_1} \mathbb{T}_{t_2}$
3. For all $\hat{x} \in X$ we have

$$\lim_{t \to 0+} \|\mathbb{T}_t \hat{x} - \hat{x}\|_X = 0.$$

4. For all $t \in [0, \infty)$ and all $\hat{x} \in X$ we have $\|\mathbb{T}_t \hat{x}\|_X \leq \|\hat{x}\|_X$.

The fact that the closure of A generates a strongly continuous semigroup of contractions follows from the LUMER–PHILLIPS THEOREM, that is stated in the language of functional analysis, see [58].

Lumer–Phillips Theorem [58] *Let X be a Banach space and $(A, D(A))$ a densely defined operator. If A and A^* are dissipative, the closure of A generates a strongly continuous semigroup of contractions.*

An excellent exposition of the use of semigroups in control is given in [54]. Using the semigroup $\mathbb{T}_t$ we can write the solutions of the initial value problem

$$\begin{cases} y(0, x) = y_0(x), \ x \in (0, L) \\ \partial_t y + \partial_x y + \partial_{xxx} y = 0 \\ y(t, 0) = y(t, L) = \partial_x y(t, L) = 0 \end{cases} \tag{6.5}$$

in the form

$$\mathbb{T}_t y_0.$$

In fact, the following theorem holds.

Theorem 6.1 ([9] Prop. 11, Prop. 13, [12], Lemma A1). *For all $y_0 \in L^2(0, L)$ the initial value problem (6.5) has a unique solution*

$$y \in C([0, T]; D(A)) \cap C^1([0, T]; L^2(0, L)).$$

With $u \in L^2(0, T)$ and

$$\tilde{h} \in L^1((0, T), L^2(0, L))$$

the initial boundary value problem

$$\begin{cases} y(0, x) = y_0(x), \ x \in (0, L) \\ \partial_t y = A y + \tilde{h} \\ y(t, 0) = y(t, L) = 0, \ \partial_x y(t, L) = u(t) \end{cases} \tag{6.6}$$

where the spatial operator A is as in (6.4) has a solution $y \in C([0, T]; L^2(0, L)) \cap L^2((0, T); H^1(0, L))$. There exists a constant $C > 0$ such that for all y_0, $\tilde{h}$, u the following inequality holds:

$$\|y\|_{C([0, T], L^2(0,L))} + \|y\|_{L^2((0, T), H^1(0,L))}$$

$$\leq C \left(\|y_0\|^2_{L^2(0,L)} + \|\tilde{h}\|^2_{L^1((0,T), L^2(0,L))} + \|u\|^2_{L^2(0,T)} \right)^{1/2}. \tag{6.7}$$

Proof of Theorem 6.1. For the homogeneous case $\tilde{h} = 0$, $u = 0$ we get the solution of (6.5) from Proposition 2.1.5 from [54]. In order to obtain the solution for the inhomogeneous case, we consider the following reduction to the homogeneous case. We define the auxiliary function

$$\varphi(x) = -\frac{1}{L}x(L - x).$$

Then we have $\varphi(0) = \varphi(L) = 0$ and $\varphi'(L) = 1$. For

$$\psi(t, x) = \varphi(x)\, u(t)$$

this implies $\psi(t, 0) = \psi(t, L) = 0$ and $\psi_x(t, L) = u(t)$. We define

$$f = (-\psi_t + A\psi + \tilde{h})$$

and the initial value problem with an inhomogeneous differential equation and homogeneous boundary conditions

$$\begin{cases} z(0, x) = y_0(x), \ x \in (0, L) \\ \partial_t z = A z + f \\ z(t, 0) = z(t, L) = \partial_x z(t, L) = 0. \end{cases} \tag{6.8}$$

For a regular control $u \in C^2(0, T)$ with $u(0) = 0$, Duhamel's formula yields the solution

$$z(t, \cdot) = \mathbb{T}_t y_0 + \int_0^t \mathbb{T}_{t-s} f(s)\, ds.$$

We define

$$y = z + \psi.$$

Then we get

$$\begin{aligned} y_t &= z_t + \psi_t \\ &= Az + f + \psi_t \\ &= Az + A\psi - \psi_t + \tilde{h} + \psi_t \\ &= A(z + \psi) + \tilde{h} \\ &= A y + \tilde{h}. \end{aligned}$$

Moreover y satisfies the inhomogeneous boundary conditions. Thus y solves the initial value problem for $u \in C^2(0, T)$. By a density argument this implies the desired result for all $u \in L^2(0, T)$. Inequality (6.7) follows by multiplication of the pde with special test function and integration by parts (see [9]). Thus we have proved Theorem 6.1. $\square$

6.1.2   A traveling waves solution for the linearized system

In order to get a better understanding of the linearized KdV-equation, we consider a traveling waves solution. In this case, the solution is the sum of three traveling waves.

Let real numbers ω_1, ω_2, and ω_3 be given such that for all $i \in \{1, 2, 3\}$ we have

$$\omega_i^3 - \omega_i = \Lambda,$$

where $\Lambda = \omega_1 \omega_2 \omega_3$. We define the function

$$y(t, x) = [\omega_3 - \omega_2] \cos(\omega_1 x + \Lambda t)$$
$$+ [\omega_1 - \omega_3] \cos(\omega_2 x + \Lambda t)$$
$$+ [\omega_2 - \omega_1] \cos(\omega_3 x + \Lambda t).$$

Then we have the time-derivative

$$y_t(t, x) = -\Lambda [\omega_3 - \omega_2] \sin(\omega_1 x + \Lambda t)$$
$$-\Lambda [\omega_1 - \omega_3] \sin(\omega_2 x + \Lambda t)$$
$$-\Lambda [\omega_2 - \omega_1] \sin(\omega_3 x + \Lambda t).$$

Moreover, for the trigonometric function $\varphi_i(x) = \cos(\omega_i x + \Lambda t)$ we have the derivatives

$$\varphi_i'(x) = -\omega_i \sin(\omega_i x + \Lambda t),$$
$$\varphi_i''(x) = -\omega_i^2 \cos(\omega_i x + \Lambda t),$$
$$\varphi_i'''(x) = \omega_i^3 \sin(\omega_i x + \Lambda t).$$

Thus we get for all $i \in \{1, 2, 3\}$

$$\varphi_i'(x) + \varphi_i'''(x) = (-\omega_i + \omega_i^3) \sin(\omega_i x + \Lambda t) = \Lambda \sin(\Lambda t + \omega_i x).$$

This implies

$$\partial_t y(t, x) = -\partial_{xxx} y - \partial_x y,$$

hence y solves the linearized KdV equation. Moreover, for $x = 0$ we have

$$y(t, 0) = [\omega_3 - \omega_2] \cos(\Lambda\, t) + [\omega_1 - \omega_3] \cos(\Lambda\, t) + [\omega_2 - \omega_1] \cos(\Lambda\, t) = 0.$$

Since for the partial derivative with respect to x we have

$$y_x(t, x) = -\omega_1 [\omega_3 - \omega_2] \sin(\omega_1 x + \Lambda\, t)$$
$$-\omega_2 [\omega_1 - \omega_3] \sin(\omega_2 x + \Lambda\, t) - \omega_3 [\omega_2 - \omega_1] \sin(\omega_3 x + \Lambda\, t),$$

we also have $y_x(t, 0) = 0$.

6.1.3 Well-posedness for the nonlinear system

In this section we examine the well-posedness for the nonlinear system. In a first step we consider the nonlinear term $y\, \partial_x y$ and show that it can be considered as a source term in the linear equation.

Lemma 6.1 (See [47], Proposition 4.1). *For all $y \in L^2((0, T), H^1(0, L))$ we have $y\, y_x \in L^1((0, T), L^2(0, L))$ and the mapping $y \mapsto y\, y_x$ is continuous.*

Proof of Lemma 6.1. Let $y, z \in L^2((0, T), H^1(0, L))$ be given. Let K denote the norm of the embedding of $H^1(0, L)$ in $L^\infty(0, L)$. Then we have the inequality

$$\|y\, y_x - z\, z_x\|_{L^1((0,T),L^2(0,L))} \leq \int_0^T \|(y - z)y_x\|_{L^2(0,L)}\, dt$$
$$+ \int_0^T \|z(y_x - z_x)\|_{L^2(0,L)}\, dt$$
$$\leq \int_0^T \|y - z\|_{L^\infty(0,L)} \|y_x\|_{L^2(0,L)}\, dt$$
$$+ \int_0^T \|z\|_{L^\infty(0,L)} \|y_x - z_x\|_{L^2(0,L)}\, dt$$
$$\leq K \int_0^T \|y - z\|_{H^1(0,L)} \|y\|_{H^1(0,L)}\, dt$$
$$+ K \int_0^T \|z\|_{H^1(0,L)} \|y_x - z_x\|_{L^2(0,L)}\, dt$$
$$\leq K \|y - z\|_{L^2((0,T),H^1(0,L))} \|y\|_{L^2((0,T),H^1(0,L))}$$
$$+ K \|z\|_{L^2((0,T),H^1(0,L))} \|y - z\|_{L^2((0,T),H^1(0,L))}\, dt$$
$$= K \left(\|y\|_{L^2((0,T),H^1(0,L))} + \|z\|_{L^2((0,T),H^1(0,L))} \right)$$
$$\|y - z\|_{L^2((0,T),H^1(0,L))}.$$

If we insert $z = 0$, we see that we have $yy_x \in L^1((0,T), L^2(0,L))$. Moreover the inequality implies the continuity of the map $y \mapsto yy_x$. This finishes the proof of Lemma 6.1. $\square$

Now we show the well-posedness of the initial boundary value problem

$$
\begin{cases}
y(0,x) = y_0(x), \ x \in (0,L) \\
\partial_t y = -\partial_x y - \partial_{xxx} y - y\,\partial_x y + \tilde{h} \\
y(t,0) = y(t,L) = 0, \ \partial_x y(t,L) = u(t).
\end{cases}
\tag{6.9}
$$

Lemma 6.2 ([11], Proposition 14). *Let $L > 0$ and $T > 0$ be given. Define the set*

$$
\mathcal{B} = C([0,T], L^2(0,L)) \cap L^2((0,T), H^1(0,L)).
$$

There exist numbers $\varepsilon > 0$ and $C > 0$ such that for all $\tilde{h} \in L^1((0,T), L^2(0,L))$, all $u \in L^2(0,T)$ and all $y_0 \in L^2(0,L)$ with

$$
\|\tilde{h}\|_{L^1((0,T),L^2(0,L))} + \|u\|_{L^2(0,T)} + \|y_0\|_{L^2(0,L)} \le \varepsilon
$$

the initial boundary value problem (6.9) has a solution y with

$$
\|y\|_{\mathcal{B}} \le C \left(\|\tilde{h}\|_{L^1((0,T),L^2(0,L))} + \|u\|_{L^2(0,T)} + \|y_0\|_{L^2(0,L)} \right).
\tag{6.10}
$$

Proof of Lemma 6.2. For $z \in \mathcal{B}$ we define the map

$$
M : \mathcal{B} \to \mathcal{B}
$$

by $M(z) = \tilde{y}$ where $\tilde{y}$ denotes the solution of the initial value problem

$$
\begin{cases}
\tilde{y}(0,x) = y_0(x), \ x \in (0,L) \\
\partial_t \tilde{y} = -\partial_x \tilde{y} - \partial_{xxx} \tilde{y} - z\,\partial_x z + \tilde{h} \\
\tilde{y}(t,0) = \tilde{y}(t,L) = 0, \ \partial_x \tilde{y}(t,L) = u(t).
\end{cases}
\tag{6.11}
$$

Theorem 6.1 implies the existence of a solution $\tilde{y}$ that satisfies (6.7). We are looking for a fixed point y of M. For this purpose, we want to apply BANACH's fixed point theorem. Inequality (6.7) and Lemma 6.1 imply that there exists a number $D > 0$ such that

$$
\|M(z)\|_{\mathcal{B}} \le D \left[\|\tilde{h}\|_{L^1((0,T),L^2(0,L))} + \|u\|_{L^2(0,T)} + \|y_0\|_{L^2(0,L)} + \|z\|_{\mathcal{B}}^2 \right].
\tag{6.12}
$$

Moreover we have

$$
\|M(z_1) - M(z_2)\|_{\mathcal{B}} \le D \left(\|z_1\|_{\mathcal{B}} + \|z_2\|_{\mathcal{B}} \right) \|z_1 - z_2\|_{\mathcal{B}}.
$$

Now we choose numbers $R > 0$ and $\varepsilon > 0$ sufficiently small such that

$$
R < \frac{1}{2D} \ \text{ and } \ \varepsilon < \frac{R}{2D}.
$$

We consider the set

$$Z = \{z \in \mathcal{B} : \|z\|_{\mathcal{B}} \leq R\}.$$

Then $M(Z) \subset Z$ since for all $z \in Z$ we have the inequality

$$\|M(z)\|_{\mathcal{B}} \leq D\left[\varepsilon + R^2\right] \leq \frac{R}{2} + DR^2 = \frac{R}{2} + (DR)R$$

$$\leq \frac{R}{2} + \frac{R}{2} = R.$$

The map M is a contraction in Z with the Lipschitz constant

$$L_M = 2DR < 1.$$

Now BANACH'S fixed point theorem implies the existence of a unique fixed point $z \in Z$ with $M(z) = z$. Since $D\|z\|_{\mathcal{B}} \leq DR < 1/2$ inserting the fixed point in (6.12) yields (6.10) with $C = 2D$. Thus Lemma 6.2 is proved. $\square$

6.1.4 A traveling wave solution for the nonlinear system

Also for the nonlinear system there is a traveling wave solution that satisfies the homogeneous KdV equation. We make the ansatz

$$y(t, x) = \frac{\alpha}{\cosh^2(ax + bt)}. \tag{6.13}$$

Then we get the partial derivatives

$$y_t(t, x) = -2\alpha\frac{\sinh(ax + bt)}{\cosh^3(ax + bt)}b, \tag{6.14}$$

$$y_x(t, x) = -2\alpha\frac{\sinh(ax + bt)}{\cosh^3(ax + bt)}a, \tag{6.15}$$

$$y_{xx}(t, x) = 6\alpha\frac{\sinh^2(ax + bt)}{\cosh^4(ax + bt)}a^2 - 2\alpha\frac{1}{\cosh^2(ax + bt)}a^2, \tag{6.16}$$

$$y_{xxx}(t, x) = -24\alpha\frac{\sinh^3(ax + bt)}{\cosh^5(ax + bt)}a^3 \tag{6.17}$$

$$+ 12\alpha\frac{\sinh(ax + bt)\,\cosh(ax + bt)}{\cosh^4(ax + bt)}a^3 \tag{6.18}$$

$$+ 4\alpha\frac{\sinh(ax + bt)}{\cosh^3(ax + bt)}a^3. \tag{6.19}$$

Since $\sinh^2(z) = \cosh^2(z) - 1$ this implies

$$y_{xxx}(t,\,x) = 24\alpha\frac{\sinh(ax+bt)}{\cosh^5(ax+bt)}a^3 - 24\alpha\frac{\sinh(ax+bt)}{\cosh^3(ax+bt)}a^3 \qquad (6.20)$$

$$+ 12\alpha\frac{\sinh(ax+bt)}{\cosh^3(ax+bt)}a^3 + 4\alpha\frac{\sinh(ax+bt)}{\cosh^3(ax+bt)}a^3 \qquad (6.21)$$

$$= 24\alpha\frac{\sinh(ax+bt)}{\cosh^5(ax+bt)}a^3 - 8\alpha\frac{\sinh(ax+bt)}{\cosh^3(ax+bt)}a^3. \qquad (6.22)$$

This yields the product

$$y\,y_x = -2\alpha^2\frac{\sinh(ax+bt)}{\cosh^5(ax+bt)}a.$$

Thus we get

$$y_x + y_{xxx} + y\,y_x - y_t \qquad (6.23)$$

$$= \left[-2\alpha\,a - 8\alpha\,a^3 + 2\alpha\,b\right]\frac{\sinh(ax+bt)}{\cosh^3(ax+bt)} \qquad (6.24)$$

$$+ \left[-2\alpha^2\,a + 24\alpha\,a^3\right]\frac{\sinh(ax+bt)}{\cosh^5(ax+bt)}. \qquad (6.25)$$

With the choice

$$\alpha = 12\,a^2 \text{ und } b = a(1 + 4\,a^2)$$

we obtain the traveling waves solution

$$y(t,\,x) = \frac{12\,a^2}{\cosh^2(a\,(x + (1 + 4a^2)t))}$$

for (6.2). The speed $(1 + 4a^2)$ can become arbitrarily large and fixes the height of the wave.

Remark 6.1. For the viscous Burgers equation

$$y_t = y_x + y\,y_x + \nu\,y_{xx}$$

with viscosity $\nu > 0$ there also exists a traveling wave solution that has the form

$$y(t,x) = \alpha\,\tanh(ax+bt). \qquad (6.26)$$

Exercise 6.1. Show that the ansatz (6.26) yields a traveling wave solution of the viscous Burgers equation with the choice $\alpha = 2\,a\,\nu$ und $b = a$. Thus for $\nu > 0$ and $a = 1/\nu$ we get the solution $y(t, x) = 2\tanh((x+t)/\nu)$. Note that the limit function for $\nu \to 0+$ is not continuous.

Remark 6.2. A detailed discussion of interacting solitary waves that solve the KdV equation can be found in [56].

6.1.5 The linearized system with critical length: An example for a system that is not exactly controllable

In this section we study the linearized system for a special choice of the length L and show that the system is **not** exactly controllable. This result is due to L. Rosier, [47]. We are looking for a length L, for which the overdetermined system

$$\begin{cases} \partial_t y = -\partial_x y - \partial_{xxx} y \\ y(t,\, 0) = y(t,\, L) = 0,\ \partial_x y(t,\, 0) = \partial_x y(t,\, L) = 0 \end{cases} \tag{6.27}$$

has a nontrivial solution. For this purpose we consider the traveling wave solution from Section 6.1.2. As an example, let

$$\omega_1 = -\frac{1}{\sqrt{7}},\ \omega_2 = -\frac{2}{\sqrt{7}},\ \omega_3 = \frac{3}{\sqrt{7}}.$$

Then we have $\Lambda = \omega_1 \omega_2 \omega_2 = \frac{6}{7\sqrt{7}}$ and

$$(\omega - \omega_1)(\omega - \omega_2)(\omega - \omega_3)$$
$$= \omega^3 - [\omega_1 + \omega_2 + \omega_3]\,\omega^2 + [\omega_1\omega_2 + \omega_2\omega_3 + \omega_1\omega_3]\,\omega - \Lambda$$
$$= \omega^3 - \omega - \frac{6}{7\sqrt{7}}.$$

We define

$$L = 2\pi\sqrt{7}.$$

Then we have

$$y(t, L) = [\omega_3 - \omega_2]\,\cos(\omega_1 L + \Lambda\, t)$$
$$+ [\omega_1 - \omega_3]\,\cos(\omega_2 L + \Lambda\, t)$$
$$+ [\omega_2 - \omega_1]\,\cos(\omega_3 L + \Lambda\, t)$$
$$= [\omega_3 - \omega_2]\,\cos(-2\pi + \Lambda\, t)$$
$$+ [\omega_1 - \omega_3]\,\cos(-4\pi + \Lambda\, t)$$
$$+ [\omega_2 - \omega_1]\,\cos(6\pi + \Lambda\, t)$$
$$= y(t, 0)$$
$$= 0.$$

Moreover we have

$$
\begin{aligned}
y_x(t, L) &= -\omega_1[\omega_3 - \omega_2] \, \sin(\omega_1 L + \Lambda t) \\
&\quad - \omega_2[\omega_1 - \omega_3] \sin(\omega_2 L + \Lambda t) \\
&\quad - \omega_3[\omega_2 - \omega_1] \sin(\omega_3 L + \Lambda t) \\
&= -\omega_1[\omega_3 - \omega_2] \, \sin(-2\pi + \Lambda t) \\
&\quad - \omega_2[\omega_1 - \omega_3] \sin(-4\pi + \Lambda t) \\
&\quad - \omega_3[\omega_2 - \omega_1] \sin(-6\pi + \Lambda t) \\
&= y_x(t, 0) = 0.
\end{aligned}
$$

Hence for $L = 2\pi \sqrt{7}$ the function y solves our system with the overdetermined boundary conditions. Figure 6.1 shows the solution on the time interval $[0, 30]$.

Remark 6.3. A key fact for the generalization of this example is the equation

$$
L = 2\pi \sqrt{7} = 2\pi \sqrt{\frac{1^2 + 4^2 + 1 \cdot 4}{3}}.
$$

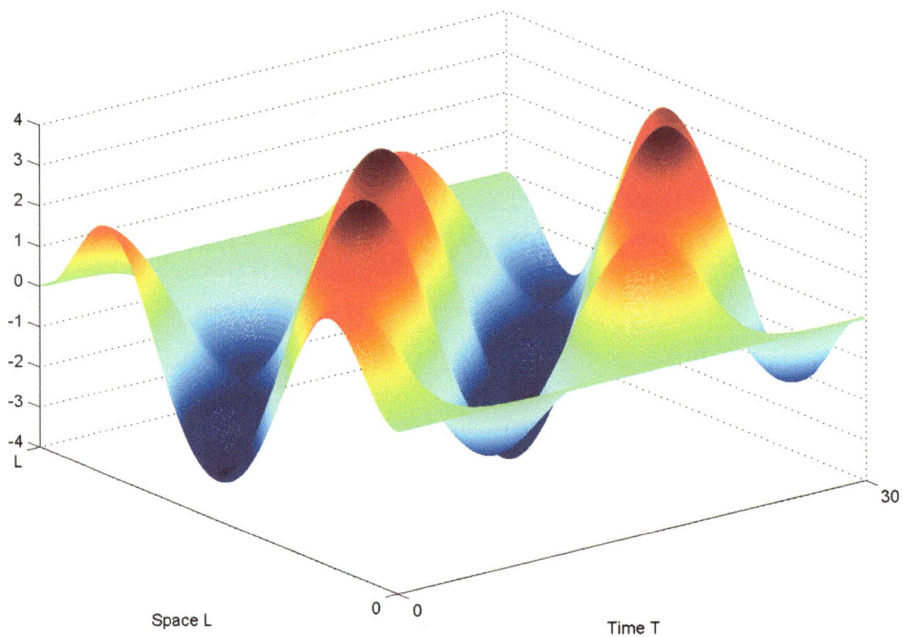

Fig. 6.1 A nontrivial traveling waves solution of the system (6.27) with overdetermined boundary conditions

In general nontrivial solutions of (6.27) exist for lengths

$$L \in \mathcal{N} = \left\{ 2\pi \sqrt{\frac{j^2 + l^2 + jl}{3}} : j, l \in \{1, 2, 3, \ldots\} \right\}$$

(see [47]). In particular, such a solution can only exist, if L is sufficiently large, that is if $L \geq 2\pi$!

The solutions of (6.27) are invisible for an observer in the following sense: If the value $y_x(t, 0)$ is observed, then the observation is always zero, no matter how long it is observed (with zero control).

Exercise 6.2. Consider the length $L = 2\sqrt{13}\,\pi$. Determine a nontrivial solution for the system (6.27) with overdetermined boundary conditions.

Solution of Exercise 6.2. *We put* $\omega_1 = -\frac{1}{\sqrt{13}}$, $\omega_2 = -\frac{3}{\sqrt{13}}$, $\omega_2 = \frac{4}{\sqrt{13}}$.

Exercise 6.3. Define the set of numbers

$$\mathcal{M} = \left\{ 2\pi \sqrt{\frac{k_2^2 - k_2 k_3 + k_3^2}{3}}, \ k_2, k_3 \in \mathbb{Z}, k_2 \neq k_3, k_2 k_3 \neq 0 \right\}.$$

Show that for all $L \in \mathcal{M}$ there is a nontrivial solution for the homogeneous system (6.27) with overdetermined boundary conditions. Use solutions of the type defined in Section 6.1.2.

Solution of Exercise 6.3. *Let the numbers* $k_2, k_3 \in \mathbb{Z}, k_2 \neq k_3, k_2 k_3 \neq 0$ *be given. We define*

$$\omega_1 = -\frac{1}{L} \frac{2\pi}{3} (k_2 + k_3),$$

$$\omega_2 = -\frac{1}{L} \frac{2\pi}{3} (-2k_2 + k_3),$$

$$\omega_3 = -\frac{1}{L} \frac{2\pi}{3} (k_2 - 2k_3).$$

Then we have $\omega_2 \neq \omega_3$, $\omega_1 \neq \omega_2$, *and* $\omega_1 \neq \omega_3$.
Hence the solution from Section 6.1.2 is nontrivial.
We have $\omega_1 + \omega_2 + \omega_3 = 0$. *Moreover*

$$\omega_1 \omega_2 + \omega_1 \omega_3 + \omega_2 \omega_3 = \frac{4\pi^2}{9L^2} [(k_2 + k_3)(-k_2 - k_3) + (-2k_2 + k_3)(k_2 - 2k_3)]$$

$$= \frac{4\pi^2}{3L^2} \left[-k_2^2 + k_2 k_3 - k_3^2 \right].$$

Hence for all $L \in \mathcal{M}$ on account of

$$L^2 = 4\pi^2 \frac{k_2^2 - k_2 k_3 + k_3^2}{3}.$$

we have the equation $\omega_1 \omega_2 + \omega_1 \omega_3 + \omega_2 \omega_3 = -1$. Thus we have

$$(\omega - \omega_1)(\omega - \omega_2)(\omega - \omega_3) = \omega^3 - \omega - \omega_1 \omega_2 \omega_3,$$

as required in Section 6.1.2. Thus we have $y(t, 0) = 0 = y_x(t, 0)$. Moreover

$$\begin{aligned}
y(t, L) &= [\omega_3 - \omega_2] \cos(\omega_1 L + \Lambda\, t) \\
&+ [\omega_1 - \omega_3] \cos(\omega_2 L + \Lambda\, t) \\
&+ [\omega_2 - \omega_1] \cos(\omega_3 L + \Lambda\, t) \\
&= [\omega_3 - \omega_2] \cos(\omega_1 L + \Lambda\, t) \\
&+ [\omega_1 - \omega_3] \cos(\omega_1 L + \Lambda\, t + 2\,\pi\, k_2) \\
&+ [\omega_2 - \omega_1] \cos(\omega_1 L + \Lambda\, t + 2\,\pi\, k_3) \\
&= 0
\end{aligned}$$

and

$$\begin{aligned}
y_x(t, L) &= -\omega_1 [\omega_3 - \omega_2]\ \sin(\omega_1 L + \Lambda\, t) \\
&- \omega_2 [\omega_1 - \omega_3] \sin(\omega_2 L + \Lambda\, t) \\
&- \omega_3 [\omega_2 - \omega_1] \sin(\omega_3 L + \Lambda\, t) \\
&= -\omega_1 [\omega_3 - \omega_2]\ \sin(\omega_1 L + \Lambda\, t) \\
&- \omega_2 [\omega_1 - \omega_3] \sin(\omega_2 L + \Lambda\, t + 2\,\pi\, k_2) \\
&- \omega_3 [\omega_2 - \omega_1] \sin(\omega_3 L + \Lambda\, t + 2\,\pi\, k_3) \\
&= 0.
\end{aligned}$$

Therefore $y(t, x)$ is a nontrivial solution for (6.27).

Exercise 6.4. Show that for the set $\mathcal{M}$ that is defined in Exercise 6.3 we have

$$\mathcal{M} = \mathcal{N}.$$

Use the equation

$$k_2^2 - k_2 k_3 + k_3^2 = (k_2 - k_3)^2 + (k_2 - k_3)k_3 + k_3^2.$$

In order to analyze the exact controllability, we consider the linear operator

$$\mathcal{F}_T : L^2(0, T) \rightarrow L^2(0, L),$$

that maps a control $u \in L^2(0, T)$ to the corresponding terminal state $y(T, \cdot) \in L^2(0, L)$ that is generated starting from the initial state $y_0 = 0$. The System (6.6) with $\tilde{h} = 0$ is exactly controllable if and only if $\mathcal{F}_T$ is surjective. To analyze the surjectivity of $\mathcal{F}_T$ we use a theorem from functional analysis (see [58]), where the adjoint map $\mathcal{F}^*$ plays a central role. Therefore, let us look at $\mathcal{F}_T^*$.

Lemma 6.3 (Lemma 2.28 in [10]). *Let $z^T \in D(A^*)$ be given. Then we have*

$$\mathcal{F}_T^*(z^T) = z_x(\cdot, L),$$

where $z \in C([0, T], H^3(0, L))$ solves the following problem:

$$\begin{cases} z(T, x) = z^T(x),\ x \in (0, L) \\ \partial_t z = -A^* z, \\ z(t, \cdot) \in D(A^*). \end{cases} \tag{6.28}$$

Proof of Lemma 6.3. Let $u \in C^2([0, T])$ with $u(0) = 0$ and the solution y of the initial boundary value problem

$$\begin{cases} y(0, x) = 0,\ x \in (0, L) \\ \partial_t y = A y \\ y(t, 0) = y(t, L) = 0,\ \partial_x y(t, L) = u(t) \end{cases} \tag{6.29}$$

be given. Then by the definition of $\mathcal{F}_T$ we have

$$\mathcal{F}_T(u) = y(T, \cdot).$$

Integration by parts implies

$$\int_0^L z^T(x)\, \mathcal{F}_T(u)(x)\, dx$$

$$= = \int_0^L z^T(x)\, y(T, x)\, dx - \int_0^L z(0, x)\, y(0, x)\, dx$$

$$= \int_0^T \int_0^L \partial_t(zy)\, dx\, dt$$

$$= \int_0^T \int_0^L (-A^* z)y + z(Ay)\, dx\, dt$$

$$= \int_0^T \int_0^L -\partial_{xxx}z\,y - z\,\partial_{xxx}y - \partial_x z\,y - z\,\partial_x y\,dx\,dt$$

$$= \int_0^T \int_0^L -\partial_{xxx}z\,y - z\,\partial_{xxx}y\;dx\,dt$$

$$= \int_0^T \int_0^L \partial_{xx}z\,\partial_x y + \partial_x z\,\partial_{xx}y\;dx\,dt$$

$$- \int_0^T \partial_{xx}z\,y|_{x=0}^{L}\,dt - \int_0^T z\,\partial_{xx}y|_{x=0}^{L}\,dt$$

$$= \int_0^T \int_0^L -\partial_x z\,\partial_{xx}y + \partial_x z\,\partial_{xx}y\;dx\,dt$$

$$- \int_0^T \partial_{xx}z\,y|_{x=0}^{L}\,dt - \int_0^T z\,\partial_{xx}y|_{x=0}^{L}\,dt + \int_0^T \partial_x z\,\partial_x y|_{x=0}^{L}\,dt$$

With the boundary conditions this yields

$$\int_0^L z^T(x)\,\mathcal{F}_T(u)(x)\,dx$$

$$= \int_0^T \partial_x z\,\partial_x y|_{x=0}^{L}\,dt - \int_0^T \partial_{xx}z\,y|_{x=0}^{L}\,dt - \int_0^T z\,\partial_{xx}y|_{x=0}^{L}\,dt$$

$$= \int_0^T \partial_x z(t,L)\,u(t)\,dt$$

and the assertion follows. Thus we have proved Lemma 6.3. $\square$

Now we can apply the following general result about the surjectivity of $\mathcal{F}$.

Theorem 6.2 (Closed Range Theorem). *Let H_1 and H_2 be Hilbert spaces and $\mathcal{F}$ a continuous linear mapping from H_1 to H_2. Then $\mathcal{F}$ is surjective if and only if there is a constant $\kappa > 0$ such that for all $x_2 \in H_2$ we have the inequality*

$$\|\mathcal{F}^*(x_2)\|_{H_1} \geq \kappa \|x_2\|_{H_2}. \tag{6.30}$$

Inequality (6.30) is called *observability inequality*. According to Lemma 6.3 for our KdV system it has the form

$$\|\partial_x z(t,L)\|_{L^2(0,T)} \geq \kappa \|z^T\|_{L^2(0,L)}, \tag{6.31}$$

where z is the solution of (6.28).

Now we consider the traveling waves solution y from Section 6.1.2. We define $z^T(x) = y(T,x)$. For the singular lengths $L \in \mathcal{N}$ the function y solves the system (6.27) with overdetermined boundary conditions. Therefore for the singular lengths

$L \in \mathcal{N}$ the function $z(t, x) = y(t, x)$ solves (6.28). Thus we have $\partial_{xz} z(t, L) = 0$. Hence inequality (6.31) cannot hold, since the left-hand side is zero, but $z^T(x) \neq 0$. By Theorem 6.2 this implies that for the singular lengths from the set $\mathcal{N}$ the linearized KdV system is not exactly controllable.

Remark 6.4. We have seen in Section 6.1.1 that for all $w \in D(A)$ we have

$$\int_0^L w \, Aw \, dx = -\frac{1}{2}(w'(0))^2.$$

Let y denote a nontrivial solution of the overdetermined system (6.27). Then $y \in D(A^*) \cap D(A)$. Now we consider the evolution of the L^2-norm

$$E(t) = \frac{1}{2} \int_0^L (y(t, x))^2 \, dx.$$

For the time-derivative we get

$$E'(t) = \int_0^L y(t, x) \, \partial_t y(t, x) \, dx = \int_0^L y(t, x) \, Ay(t, x) \, dx = 0.$$

Hence the function E is constant, that is the L^2-norm is a conserved quantity.

Remark 6.5. For $L \notin \mathcal{N}$ the linearized KdV system is exactly controllable (see [47]).

Remark 6.6. For $L \notin \mathcal{N}$ also the nonlinear system is locally exactly controllable. Locally means that the L^2-norm of the initial and the terminal state has to be sufficiently small. In fact, the nonlinear KdV system is also exactly controllable for $L \in \mathcal{N}$, see [9].

6.2   The isothermal Euler equations

As an example for a quasilinear 2×2 system we consider the *isothermal Euler equations* that can be used as a model for the flow of gas through pipelines:

$$\begin{cases} \rho_t + q_x = 0, \\ q_t + (\frac{q^2}{\rho} + a^2 \rho)_x = -\frac{1}{2}\theta \frac{q|q|}{\rho}. \end{cases} \tag{6.32}$$

Here ρ is the density and q the flow rate of the gas. The first equation guarantees the conservation of mass. In the second equation, a is the sound speed and $\theta = \frac{f_g}{\delta}$ where δ is the diameter of the pipe and f_g is a friction parameter. To see the connection to the wave equation, let us look at the velocity

$$v = \frac{q}{\rho}.$$

For the applications the *subcritical flows* are interesting, where we have

$$|v| = \left|\frac{q}{\rho}\right| < a$$

that is the velocity is less than the sound speed. For sufficiently regular solutions v solves a *quasilinear* wave equation, namely

$$v_{tt} = (a^2 - v^2)v_{xx} - 2vv_{tx} - 2v_t v_x - 2v(v_x)^2 - \theta|v|\, v_t - \frac{3}{2}\theta v\, |v|\, v_x. \qquad (6.33)$$

Exercise 6.5. Show that the velocity v satisfies the quasilinear wave equation (6.33)!

Solution of Exercise 6.5. *Since $q = \rho\, v$, the equation $\rho_t + q_x = 0$ implies $\rho_t + \rho_x v + \rho\, v_x = 0$, hence we have*

$$\rho_t + v\rho_x = -\rho\, v_x. \qquad (6.34)$$

Since we do not consider the vacuum case, for the density we have $\rho > 0$, therefore we can write the above equation in the form

$$\frac{\rho_t}{\rho} + v\frac{\rho_x}{\rho} = -v_x. \qquad (6.35)$$

With the variable $\ln(\rho)$ this yields

$$\partial_t(\ln(\rho)) + v\,\partial_x(\ln(\rho)) = -v_x. \qquad (6.36)$$

Hence we have

$$\partial_x(\ln(\rho)) = -\frac{v_x}{v} - \frac{1}{v}\,\partial_t(\ln(\rho)). \qquad (6.37)$$

The second equation in (6.32) yields

$$v\,\rho_t + \rho v_t = -(\rho v^2 + a^2\, \rho)_x - \frac{1}{2}\theta\, v\, |v|\, \rho.$$

With (6.34) this implies

$$v(-v\rho_x - \rho\, v_x) + \rho v_t = -2\rho v v_x - \rho_x v^2 - a^2\rho_x - \frac{1}{2}\,\theta v\, |v|\, \rho.$$

Division by ρ yields

$$v_t + v\, v_x = -a^2\partial_x\ln(\rho) - \frac{1}{2}\,\theta\, v\, |v|. \qquad (6.38)$$

Replacing the term $\partial_x \ln(\rho)$ in (6.38) by the right-hand side of (6.37) yields

$$v_t + v\,v_x = \frac{a^2}{v}\,\partial_t(\ln(\rho)) + a^2\,\frac{v_x}{v} - \frac{1}{2}\,\theta\,v\,|v|. \tag{6.39}$$

By multiplication with v this implies

$$v\,v_t + v^2\,v_x = a^2\,\partial_t(\ln(\rho)) + a^2\,v_x - \frac{1}{2}\,\theta\,v^2\,|v|. \tag{6.40}$$

By partial differentiation of (6.40) with respect to x we get

$$v_x\,v_t + v\,v_{tx} + v^2\,v_{xx} + 2v(v_x)^2 = a^2\,\partial_x\,\partial_t(\ln(\rho)) + a^2\,v_{xx} - \frac{3}{2}\,\theta\,v\,|v|\,v_x. \tag{6.41}$$

By partial differentiation of (6.38) with respect to t we obtain

$$v_{tt} + v_t\,v_x + v\,v_{tx} = -a^2\partial_t\,\partial_x\ln(\rho) - \theta\,|v|\,v_t. \tag{6.42}$$

Adding (6.41) and (6.42) yields

$$v_{tt} + 2v_x\,v_t + 2v\,v_{tx} + v^2\,v_{xx} + 2v(v_x)^2 = a^2\,v_{xx} - \theta\,|v|\left(\frac{3}{2}\,v\,v_x + v_t\right). \tag{6.43}$$

Thus we get (6.33).

Let $\mathcal{I}$ denote the identity operator. We have the equation

$$
\begin{aligned}
& v_{tt} - (a^2 - v^2)v_{xx} + 2v\,v_{tx} + 2v_t\,v_x + 2v(v_x)^2 \\
=\;& [\partial_t + (a+v)\partial_x + v_x\mathcal{I}]\,[\partial_t - (a-v)\partial_x]\,v \\
=\;& [\partial_t - (a-v)\partial_x + v_x\mathcal{I}]\,[\partial_t + (a+v)\partial_x]\,v.
\end{aligned}
\tag{6.44}
$$

Exercise 6.6. Show equation (6.44)!

Thus the solutions of the equations

$$v_t - (a - v)v_x = 0, \tag{6.45}$$

$$v_t + (a + v)v_x = 0 \tag{6.46}$$

also solve the quasilinear wave equation (6.33) with $\theta = 0$, that is

$$v_{tt} - (a^2 - v^2)v_{xx} + 2\left[v\,v_{tx} + v_t\,v_x + v(v_x)^2\right] = 0. \tag{6.47}$$

Equations (6.45), (6.46) are called (nonviscous) BURGERS equations. Often these equations appear with $a = 0$.

The solutions of these equations can be determined with the method of *char-acteristics*. The characteristic curves $\xi^v(s, x, t)$ are defined as the solutions of the initial value problems

$$\xi^v(t, x, t) = x, \quad \partial_s \xi^v(s, x, t) = \pm a + v(s, \xi^v(s, x, t)). \tag{6.48}$$

Hence the $\xi^v(s, x, t)$ solve the integral equations

$$\xi^v(s, x, t) = x \pm a(s - t) + \int_t^s v(\tau, \xi^v(\tau, x, t)) \, d\tau.$$

Now we consider the values of v along the characteristic curves. For the correspond-ing auxiliary function

$$h(s) = v(s, \xi^v(s, x, t))$$

we have

$$h'(s) = v_t + v_x \partial_s \xi^v = v_t + (v \pm a) v_x.$$

For a solution v of (6.45) ((6.46) respectively) this implies $h'(s) = 0$, hence v is constant along the characteristic curves. This implies that the characteristic curves have constant slopes, hence they are straight lines. Therefore in general different characteristic curves will intersect after finite time. In this case the solution in the sense of characteristics breaks down and a *shock* develops, namely a discontinuity of the solution. Before this happens, waves that appear in the solution become steeper and steeper. It can also happen that the characteristic curves diverge. In this case, a so-called *rarefaction fan* develops. Thus we see that *classical solutions of quasilinear equations can break down after finite time.*

On the other hand, for many quasilinear systems there exist so-called *semi-global* classical solutions that have the following property: *For a given time $T > 0$ there exists a classical solution on the time interval $[0, T]$, if the initial data and the boundary data are sufficiently small with respect to the C^1-norm and satisfy the C^1-compatibility conditions at the points where both the initial conditions and the boundary conditions hold at the initial time.*

A well-written introduction to the mathematical theory of waves is given in [37]. A detailed account of controllability and observability for quasilinear hyperbolic systems in the framework of semi-global classical solutions is given in [43]. In the next section, we present a result about semi-global Lipschitz-continuous solutions.

6.3   An initial boundary value problem for the Burgers equation

Theorem 6.3 states that for Lipschitz continuous initial data and boundary data that is Lipschitz-compatible with the initial state and sufficiently small (that is with sufficiently small maximum-norm and Lipschitz constant) the initial boundary value problem (*BARWP*) for the Burgers equation has a semi-global Lipschitz-continuous solution on the time-interval $[0, T]$ for a given time $T > 0$ in the sense of characteristics.

Theorem 6.3 (Quasilinear initial boundary value problem). *Let $T > 0$ and $a \in (0, \infty)$ be given. Assume that the function v_0 is Lipschitz continuous on $[0, \infty)$ and that u is Lipschitz continuous on $[0, T]$. We consider the system*

$$(BARWP) \begin{cases} v(0, x) = v_0(x), \ x \in (0, \infty) \\ v(t, 0) = u(t), \ t \in (0, T) \\ \partial_t v \quad = -(a + v) \, \partial_x v. \end{cases}$$

Define the numbers

$$m = \inf_{(t,x) \in [0,T] \times [0,\infty)} \{u(t), \ v_0(x)\}, \ M = \sup_{(t,x) \in [0,T] \times [0,\infty)} \{u(t), \ v_0(x)\}.$$

We assume that

$$m > -a, \tag{6.49}$$

and

$$M < a. \tag{6.50}$$

We define

$$\kappa = \max\{-m, \ M\} < a.$$

Assume that the C^0-compatibility conditions between v_0 and u hold, that is $v_0(0) = u(0)$. Let

$$\tilde{L}_R$$

denote a common Lipschitz constant for u and v_0 on $[0, T] \times \{0\} \cup \{0\} \times [0, \infty)$, such that we have

$$|u(t) - v_0(x)| \leq \tilde{L}_R \, |at + x| \tag{6.51}$$

for all $t \in [0, T]$ and $x \geq 0$. Assume that

1. *$T < \frac{1}{\tilde{L}_R}$,*
2. *$T < \frac{a - \kappa}{\tilde{L}_R}$*
   and
3. *$T < \left(1 - \frac{\kappa}{a}\right) \frac{1}{\tilde{L}_R} = \frac{a - \kappa}{a \tilde{L}_R}$.*

Then there exists a solution of (BARWP) on $[0, T]$ in the sense of characteristics.

Remark 6.7. To make sure that the solution exists on a given, possibly large time interval $[0, T]$, the Lipschitz constant $\tilde{L}_R$ must be sufficiently small.

Fig. 6.2 We assume that the
joint function that consists of
u and v_0 glued together at the
corner at 0 is Lipschitz
continuous in the sense
of (6.51) around the corner
where the functions u and v_0
are glued together.

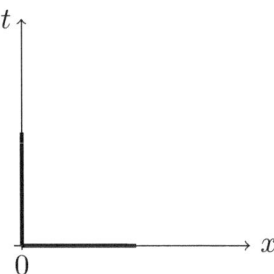

For a given value of $\tilde{L}_R$, Theorem 6.3 guarantees the existence of the solution
only on a possibly short time interval $[0, T]$.

Remark 6.8. Condition (6.51) (see also (6.57) below) means that u and v_0 are
Lipschitz continuous around the corner that is depicted in Figure 6.2.

Proof of Theorem 6.3. We consider a solution in the sense of characteristics. For the
characteristic curves we have for $x \geq 0$ the explicit representation

$$\xi(s, x, 0) = x + (a + v_0(x))\, s \tag{6.52}$$

and for $t > 0$

$$\xi(s, 0, t) = (a + u(t))\, (s - t). \tag{6.53}$$

Due to the Lipschitz-continuity of u and v_0 and the compatibility condition $v_0(0) =
u(0)$, the Picard-Lindelöf Theorem implies that the functions $\xi(\cdot, x, t)$ are uniquely
defined as the solutions of the initial value problems (6.48) for sufficiently small
$s > 0$, $t \geq 0$. Moreover, no regions in $(0, t) \times (0, \infty)$ occur, that do not contain
characteristics, that is no rarefaction fans occur.

Now we determine a time interval, where it is impossible that the characteristic
curves intersect.

In our case of the Burgers equation the *breaking time* can be estimated
quite accurately. There are three possibilities how two characteristic curves can
intersect:

1. Two characteristic curves of the type (6.52) intersect:
   The equation $\xi(s, x_1, 0) = \xi(s, x_2, 0)$ implies

$$s = -\frac{1}{\dfrac{v_0(x_2) - v_0(x_1)}{x_2 - x_1}}.$$

   Thus for a point s of intersection we have

$$|s| \geq \frac{1}{\tilde{L}_R}.$$

   Hence if $T < \frac{1}{\tilde{L}_R}$, an intersection of this kind cannot occur.

2. Two characteristic curves of the type (6.53) intersect:
   The equation $\xi(s, 0, t_1) = \xi(s, 0, t_2)$ implies

$$a = \frac{u(t_2) - u(t_1)}{t_2 - t_1} s + \frac{u(t_1) t_1 - u(t_2) t_2}{t_2 - t_1}$$

$$= s \frac{u(t_2) - u(t_1)}{t_2 - t_1} + \frac{(u(t_1) - u(t_2)) t_1 + u(t_2)(t_1 - t_2)}{t_2 - t_1}$$

$$= s \frac{u(t_2) - u(t_1)}{t_2 - t_1} - t_1 \frac{u(t_2) - u(t_1)}{t_2 - t_1} - u(t_2).$$

Hence we have

$$|s - t_1| = \left| \frac{a + u(t_2)}{\frac{u(t_2) - u(t_1)}{t_2 - t_1}} \right|$$

$$\geq \frac{a - \kappa}{\tilde{L}_R}$$

Thus if $T < \frac{a - \kappa}{\tilde{L}_R}$, an intersection of this kind cannot occur.
3. A characteristic curves of the type (6.52) intersects a characteristic curves of the type (6.53): The equation $\xi(s, 0, t) = \xi(s, x, 0)$ implies

$$s = \frac{-x - at - u(t)t}{v_0(x) - u(t)}$$

$$= \frac{1}{-\frac{v_0(x) - u(t)}{x + at}} - u(t)t \frac{1}{v_0(x) - u(t)}$$

and thus we get the inequality

$$|s| \geq \left| \frac{1}{\frac{v_0(x) - u(t)}{x + at}} - \frac{u(t)}{a}(at + x) \frac{1}{v_0(x) - u(t)} \right|$$

$$\geq \frac{1}{\tilde{L}_R} - \frac{\kappa}{a} \frac{1}{\tilde{L}_R}$$

$$= \left(1 - \frac{\kappa}{a}\right) \frac{1}{\tilde{L}_R}.$$

If $T < \left(1 - \frac{\kappa}{a}\right) \frac{1}{\tilde{L}_R}$, an intersection of this kind cannot occur.

Thus we have proved Theorem 6.3. $\square$

Remark 6.9. In order to extend the solution if a shock has developed, only a solution in a weaker sense can be chosen. However, in general the weak solutions are not uniquely determined. It makes sense to choose the solution that is obtained as a limit of the solutions of the viscous Burgers equation for vanishing viscosity (see Exercise 6.1).

6.4   The Burgers equation with source term

In the applications, often source terms appear in the balance laws, for example to model the effects of friction. Thus in this section we study the solutions of an initial boundary value problem for the Burgers equation with a source term. Let a continuous function

$$g : \mathbb{R} \to \mathbb{R}$$

with $g(0) = 0$ and $\theta, L_g \in [0, \infty)$ be given, such that for all $z \in [-a, a]$ we have:

$$|g(z)| \leq \theta |z|^2.$$

Assume that for all $z_1, z_2 \in [-a, a]$ we have the Lipschitz inequality

$$|g(z_1) - g(z_2)| \leq L_g |z_1 - z_2|.$$

Theorem 6.4 (Quasilinear initial boundary value problem with source term).
Let $T > 0$ and $a \in (0, \infty)$ be given. Assume that the function v_0 is Lipschitz continuous on $[0, \infty)$ and that u is Lipschitz continuous on $[0, T]$. We consider the system

$$(QARWP) \begin{cases} v(0, x) = v_0(x), \ x \in (0, \infty) \\ v(t, 0) = u(t), \ t \in (0, T) \\ \partial_t v \ \ = -(a + v) \, \partial_x v + g(v). \end{cases}$$

Define the numbers

$$m = \inf_{(t,x) \in [0,T] \times [0,\infty)} \{u(t), \ v_0(x)\}, \ M = \sup_{(t,x) \in [0,T] \times [0,\infty)} \{u(t), \ v_0(x)\}.$$

We assume that

$$m > -a, \tag{6.54}$$

and

$$M < a. \tag{6.55}$$

We define

$$\kappa = \max\{-m, M\} + \theta \, T \, a^2 \tag{6.56}$$

and assume that

$$\kappa < a.$$

Assume that the C^0-compatibility conditions between v_0 and u hold, that is $v_0(0) = u(0)$. Let

$$\tilde{L}_R$$

denote a common Lipschitz constant for u and v_0 on $[0, T] \times \{0\} \cup \{0\} \times [0, \infty)$, such that we have

$$|u(t) - v_0(x)| \leq \tilde{L}_R |at + x| \tag{6.57}$$

for all $t \in [0, T]$ and $x \geq 0$. Assume that there exists a number $L_M > \tilde{L}_R$, such that

$$L_{kontr} := T \exp(L_M T) \left[L_g (1 + L_M T) + \frac{\theta a^2}{a - \kappa} + \tilde{L}_R \left(1 + \frac{a}{a - \kappa} \right) \right] < 1$$

and

$$\exp(L_M T) \left[L_g L_M T + \frac{\theta a^2}{a - \kappa} + \tilde{L}_R \left(1 + \frac{a}{a - \kappa} \right) \right] \leq L_M.$$

Then there exists a solution of (QARWP) on $[0, T]$ in the sense of characteristics.

Remark 6.10. Due to the source term, in general now the characteristic curves are not given by straight lines. In contrast to the case $g = 0$ for nonvanishing source term in general the constant states are not stationary, since the solutions are not constant along the characteristic curves.

For the following proofs we use a fundamental Lemma of THOMAS HAKON GRÖNWALL, (1877–1932).

Lemma 6.4 (Gronwall's Lemma). *Let real numbers $L > 0$, $U_0 \geq 0$, $\varepsilon \geq 0$, and a continuous function U be given.*

Assume that for all $t \in [0, T]$ we have the integral inequality

$$0 \leq U(t) \leq U_0 + \int_0^t L U(\tau) + \varepsilon \, d\tau.$$

Then for all $t \in [0, T]$ for $U(t)$ we have the upper bound

$$U(t) \leq U_0 \, e^{Lt} + \varepsilon \, \frac{e^{Lt} - 1}{L}.$$

Proof of Lemma 6.4. We define the auxiliary function

$$F(t) = U_0 + \int_0^t L U(\tau) + \varepsilon \, d\tau.$$

Then we have $F'(t) = LU(t) + \varepsilon$ and $U(t) \leq F(t)$. Since $L > 0$ this implies the inequality $F'(t) \leq LF(t) + \varepsilon$.

We define $H(t) = e^{-Lt} F(t)$. Then $H(0) = F(0) = U_0$. The product rule for differentiation implies

$$
\begin{aligned}
H'(t) &= -LH(t) + e^{-Lt} F'(t) \\
&\leq -LH(t) + e^{-Lt} (LF(t) + \varepsilon) \\
&= -LH(t) + LH(t) + e^{-Lt} \varepsilon \\
&= \varepsilon e^{-Lt}.
\end{aligned}
$$

By integration, the inequality $H'(\tau) \leq \varepsilon e^{-L\tau}$ yields

$$
\begin{aligned}
H(t) - H(0) &= \int_0^t H'(\tau) \, d\tau \\
&\leq \int_0^t \varepsilon e^{-L\tau} \, d\tau \\
&= \varepsilon \frac{1}{L}(1 - e^{-Lt}).
\end{aligned}
$$

Hence we have

$$
\begin{aligned}
U(t) \leq F(t) &= e^{Lt} H(t) \\
&\leq e^{Lt} \left(H(0) + \varepsilon \frac{1}{L}(1 - e^{-Lt}) \right) \\
&= e^{Lt} U_0 + \varepsilon \frac{1}{L} \left(e^{Lt} - 1 \right).
\end{aligned}
$$

Thus we have proved Lemma 6.4. $\square$

For the proof of Theorem 6.4 we consider again the solution of our system in the sense of characteristics that is described by characteristic curves. For a given function v, the following lemma guarantees the existence of the characteristic curves without intersection on a given (possibly large) time interval $[0, T]$.

Lemma 6.5. *Let $T > 0$ be given. Let $v \in C([0, T] \times [0, \infty))$ be Lipschitz continuous with respect to x with the Lipschitz constant L_v. Assume that there is a real number $v_{\max}$ such that for all $(t, x) \in [0, T] \times [0, \infty)$*

$$
|v(t, x)| \leq v_{\max} < a.
$$

Then the characteristic curves $\xi^v(s, x, t)$ exist for all

$$
(s, x, t) \in [0, T] \times [0, \infty) \times [0, T].
$$

The functions $\xi^v(s, x, t)$ are continuously differentiable with respect to s with a Lipschitz continuous derivative (with the Lipschitz constant $a + v_{max}$).

For all $w \in C([0, T] \times [0, \infty))$ (with the Lipschitz constant L_w) we have the inequality

$$|\xi^v(s, x, t) - \xi^w(s, x, t)| \leq \|v - w\|_{C([0,T] \times [0, \infty))} \frac{\exp(L_v s) - 1}{L_v} \tag{6.58}$$

$$\leq T \exp(L_v T) \|v - w\|_{C([0,T] \times [0,\infty))}.$$

If $(t, x) \in [0, T] \times [0, \infty)$ is such that for all $s \in [0, T]$ we have $\xi^v(s, x, t) > 0$, we define $t^v(x, t) = 0$. Else we define $t^v(x, t) \in [0, T]$ as the solution of the equation

$$\xi^v(t^v(x, t), x, t) = 0.$$

Then we have the inequality

$$|t^v(x, t) - t^w(x, t)| \leq \frac{1}{a - v_{max}} T \exp(L_v T) \|v - w\|_{C([0, T] \times [0,\infty))}. \tag{6.59}$$

For all $x_1, x_2 \in [0, \infty)$, $s, t \in [0, T]$ we have

$$|\xi^v(s, x_1, t) - \xi^v(s, x_2, t)| \leq |x_1 - x_2| \exp(L_v s) \tag{6.60}$$

and

$$|t^v(x_1, t) - t^v(x_2, t)| \leq \frac{1}{a - v_{max}} \exp(L_v T) |x_1 - x_2|. \tag{6.61}$$

Proof of Lemma 6.5. We define a fixed point iteration for $\xi(s, x, t) = \xi^v(s, x, t)$. For this purpose we extend v on the whole x-axis by defining $v(t, x) = v(t, 0)$ if $x < 0$. Then the extension is continuous on $[0, T] \times \mathbb{R}$ and Lipschitz continuous with respect to x with the Lipschitz constant L_v. Now we consider the integral equation

$$\xi^v(s, x, t) = x + a(s - t) + \int_t^s v(\tau, \xi^v(\tau, x, t)) \, d\tau \tag{6.62}$$

for $(s, x, t) \in [0, T] \times \mathbb{R} \times [0, T]$. In order to show the existence of a unique solution we consider the corresponding PICARD-LINDELÖF iteration with the starting point

$$\xi^{(1)}(s, x, t) = x + a(s - t)$$

and

$$\xi^{(k+1)}(s, x, t) = x + a(s - t) + \int_t^s v(\tau, \xi^{(k)}(\tau, x, t)) \, d\tau$$

for $k \in \{1, 2, 3, \ldots\}$. For all $k \in \{2, 3, 4 \ldots\}$ we have the inequality

$$\left| \xi^{(k+1)}(s, x, t) - \xi^{(k)}(s, x, t) \right|$$

$$= \left| \int_t^s v(\tau, \xi^{(k)}(\tau, x, t)) - v(\tau, \xi^{(k-1)}(\tau, x, t)) \, d\tau \right|$$

$$\leq L_v \left| \int_t^s |\xi^{(k)}(\tau, x, t) - \xi^{(k-1)}(\tau, x, t)| \, d\tau \right|.$$

We have

$$\left| \xi^{(2)}(s, x, t) - \xi^{(1)}(s, x, t) \right| = \left| \int_t^s v(\tau, \xi^{(1)}(\tau, x, t)) \, d\tau \right|$$

$$\leq v_{\max} |t - s|.$$

By induction this implies

$$\left| \xi^{(k+1)}(s, x, t) - \xi^{(k)}(s, x, t) \right| \leq \frac{1}{k!} \, v_{\max} \, L_v^{k-1} \, |t - s|^k$$

$$= \frac{v_{\max}}{L_v} \frac{1}{k!} \, (|t - s| \, L_v)^k.$$

On account of

$$\sum_{k=0}^{\infty} \frac{1}{k!} \, (|t - s| \, L_v)^k = \exp \left(|t - s| \, L_v \right) < \infty$$

this implies: The sequence $(\xi^{(k)}(s, x, t))_k$ is a Cauchy sequence in the space

$$C([0, T] \times \mathbb{R} \times [0, T])$$

and hence convergent, since we have

$$\left| \xi^{(m)}(s, x, t) - \xi^{(n)}(s, x, t) \right| \leq \sum_{k=n}^{m-1} \left| \xi^{(k+1)}(s, x, t) - \xi^{(k)}(s, x, t) \right|$$

$$\leq \frac{v_{\max}}{L_v} \left| \sum_{k=n}^{m-1} \frac{1}{k!} \, (|t - s| \, L_v)^k \right|$$

$$\to 0 \text{ for } m, n \to \infty.$$

Hence there exists a limit function $\xi(s, x, t) \in C([0, T] \times \mathbb{R} \times [0, T])$ that satisfies the integral equation (6.62). Now let $\psi(s, x, t)$ be an arbitrary solution of the integral equation (6.62). Then we have

$$|\xi(s, x, t) - \psi(s, x, t)| \le \left| \int_t^s v(\tau, \xi(s, x, \tau)) - v(\tau, \psi(s, x, \tau)) \, d\tau \right|$$

$$\le \left| \int_t^s 2 v_{\max} \, d\tau \right|$$

$$\le 2 |t - s| v_{\max}.$$

Hence for all $s, t \in [0, T]$, $x \in \mathbb{R}$ we have

$$|\xi(s, x, t) - \psi(s, x, t)| \le 2 v_{\max} |t - s|.$$

By induction this implies

$$|\xi(s, x, t) - \psi(s, x, t)| \le 2 \frac{v_{\max}}{L_v} \frac{1}{k!} L_v^k |t - s|^k$$

which yields

$$\sup_{s, t \in [0,T], x \in \mathbb{R}} |\xi(s, x, t) - \psi(s, x, t)| \le 2 \frac{v_{\max}}{L_v} \frac{1}{k!} (L_v T)^k \to 0 \ (k \to \infty).$$

Hence the solution of the integral equation (6.62) is uniquely determined.

Now we show that (6.58) holds. The integral equation (6.62) implies

$$|\xi^v(s, x, t) - \xi^w(s, x, t)|$$

$$= \left| \int_t^s v(\tau, \xi^v(\tau, x, t)) - w(\tau, \xi^w(\tau, x, t)) \, d\tau \right|$$

$$\le \left| \int_t^s v(\tau, \xi^v(\tau, x, t)) - v(\tau, \xi^w(\tau, x, t)) \, d\tau \right|$$

$$+ \left| \int_t^s v(\tau, \xi^w(\tau, x, t)) - w(\tau, \xi^w(\tau, x, t)) \, d\tau \right|$$

$$\le L_v \left| \int_t^s |\xi^v(\tau, x, t) - \xi^w(\tau, x, t)| \, d\tau \right|$$

$$+ |t - s| \max_{(\tau, z) \in [0,T] \times [0,\infty)} |v(\tau, z) - w(\tau, z)|.$$

Now we can apply Lemma 6.4. We define

$$U(s) = |\xi^v(s, x, t) - \xi^w(s, x, t)|.$$

Then we have the integral inequality

$$U(s) \leq \left| \int_t^s L_v \, U(\tau) + \|v - w\|_{C([0,T] \times [0,\infty))} \, d\tau \right|.$$

Gronwall's Lemma (Lemma 6.4) yields

$$U(s) \leq \|v - w\|_{C([0,T] \times [0,\infty))} \, \frac{\exp(L_v s) - 1}{L_v}.$$

Hence we have shown (6.58). Now we show (6.59). Without restriction of generality we assume that

$$t^w(x,t) > t^v(x,t) \geq 0.$$

We have

$$\xi^w(t^w(x,t),\, x,\, t) - \xi^w(t^v(x,t),\, x,\, t) = \int_{t^v(x,t)}^{t^w(x,t)} \partial_s \xi^w(s,\, x,\, t) \, ds$$

$$= \int_{t^v(x,t)}^{t^w(x,t)} a + w(s,\, \xi^w(s,\, x,\, t) \, ds$$

$$\geq (a - w_{\max}) \left[t^w(x,t) - t^v(x,t) \right].$$

Moreover we have

$$\xi^w(t^w(x,t),\, x,\, t) - \xi^w(t^v(x,t),\, x,\, t)$$
$$= 0 - \xi^w(t^v(x,t),\, x,\, t)$$
$$\leq \xi^v(t^v(x,t),\, x,\, t) - \xi^w(t^v(x,t),\, x,\, t)$$
$$\leq T \exp(L_w T) \|v - w\|_{C([0,T] \times [0,\infty))},$$

where the last inequality follows from (6.58). Putting the last two inequalities together yields (6.59). Now Gronwall's Lemma also yields (6.60).

Exercise 6.7. Show that (6.60) holds.

Now we show (6.61). Let $x_1, x_2 \in [0, \infty)$ be given. Without loss of generality we assume that $x_1 < x_2$. Then we have

$$t^v(x_1,t) \geq t^v(x_2,t).$$

Case 1: If $t^v(x_1,t) = 0$, then $t^v(x_2,t) = 0$ and thus $t^v(x_1,t) - t^v(x_2,t) = 0$.
Case 2: If $t^v(x_1,t) > 0$, we have

$$\xi^v(t^v(x_1,t),\, x_1,\, t) = 0.$$

This implies

$$\xi^v(t^v(x_1,t),\, x_1,\, t) - \xi^v(t^v(x_2,t),\, x_1,\, t) = \int_{t^v(x_2,t)}^{t^v(x_1,t)} \partial_s \xi^v(s,\, x_1,\, t)\, ds$$

$$\geq (a - v_{\max})\, [t^v(x_1,t) - t^v(x_2,t)]\,.$$

On the other hand we have

$$\xi^v(t_+^u(x_1,t),\, x_1,\, t) - \xi^v(t^v(x_2,t),\, x_1,\, t)$$

$$= 0 - \xi^v(t^v(x_2,t),\, x_1,\, t)$$

$$\leq \xi^v(t^v(x_2,t),\, x_2,\, t) - \xi^v(t^v(x_2,t),\, x_1,\, t) \leq \exp(L_v T)\, |x_1 - x_2|$$

where the last inequality follows from (6.60). Thus (6.61) holds and we have shown Lemma 6.5. $\Box$

Proof of Theorem 6.4. We use a fixed point argument to show the existence of the solution v. Assume that L_M is as in Theorem 6.4 and define the set

$$M = \{v : v \text{ is continuous and Lipschitz continuous with respect to } x \text{ on } [0, T] \times [0, \infty)$$

$$\text{with a Lipschitz constant } L_v \leq L_M \text{ and } |v| \leq \kappa\}$$

where κ is as in (6.56). Now we consider a mapping Φ that is defined on the set M. For a given function $v \in M$ we have $v \in C([0, T] \times [0, \infty))$ with the Lipschitz constant $L_v \leq L_M$ and we have

$$|v(t, x)| \leq \kappa < a.$$

Now Lemma 6.5 implies the existence of characteristic curves ξ^v. We use these characteristic curves for the definition of the map Φ. We define

$$\Phi(v)(t, x) = v(t^v(x, t),\, \xi^v(t^v(x, t),\, x,\, t))$$

$$+ \int_{t^v(x,t)}^{t} g\left(v(\tau,\, \xi^v(\tau,\, x,\, t))\right) d\tau.$$

Note that we have

$$v(t^v(x, t),\, \xi^v(t^v(x, t),\, x,\, t)) = \begin{cases} u(t^v(x, t)) & \text{if } t^v(x, t) > 0, \\ v_0(\xi^v(0,\, x,\, t)) & \text{if } t^v(x, t) = 0 \end{cases}$$

hence the corresponding values are determined by u and v_0.

Now we consider the fixed point iteration that is defined by the equation

$$v_{k+1}(t, x) = v_k(t^{v_k}(x, t), \xi^{v_k}(t^{v_k}(x, t), x, t))$$

$$+ \int_{t^{v_k}(x,t)}^{t} g\left(v_k(\tau, \xi^{v_k}(\tau, x, t))\right) d\tau$$

$$= \Phi(v_k)(t, x).$$

Our aim is to apply Banach's fixed point theorem. We divide the corresponding analysis in 3 steps.

Step 1: The velocity remains subcritical For all $v_1 \in M$ we have

$$|v_1(t, x)| \le \max\{-m, M\} + \theta T a^2 = \kappa < a.$$

By induction this implies that for all $k \in \{0, 1, 2, \ldots\}$ we have

$$|v_{k+1}(t, x)| \le \kappa < a. \tag{6.63}$$

By (6.63) for all $k \in \{1, 2, 3, \ldots\}$ we have a subcritical flow. In particular Lemma 6.5 guarantees the existence of the characteristic curves, hence the fixed point iteration is well defined.

Let L_{v_k} denote a Lipschitz constant of v_k with respect to x.

Exercise 6.8. Show that the Lipschitz constants L_{v_k} can be chosen in such a way that the sequence $(L_{v_k})_k$ is bounded by L_M.

Step 2: The Lipschitz constants are uniformly bounded.

Now we consider the Lipschitz constants of $\Phi(v)$ with respect to x.
For all $x_1, x_2 \in [0, \infty)$ we have the inequality

$$|\Phi(v)(t, x_1) - \Phi(v)(t, x_2)| \le f_1(t, x) + f_2(t, x),$$

with

$$f_1(t, x) = |v(t^v(x_1, t), \xi^v(t^v(x_1, t), x_1, t)) - v(t^v(x_2, t), \xi^v(t^v(x_2, t), x_2, t))|$$

and

$$f_2(t, x)$$

$$= \left| \int_{t^v(x_1,t)}^{t} g(v(\tau, \xi^v(\tau(x_1, t), x_1, t)) d\tau - \int_{t^v(x_2,t)}^{t} g(v(\tau, \xi^v(\tau(x_2, t), x_2, t)) d\tau \right|$$

$$\le \left| \int_{t^v(x_1,t)}^{t} g(v(\tau, \xi^v(\tau(x_1, t), x_1, t)) - g(v(\tau, \xi^v(\tau(x_2, t), x_2, t)) d\tau \right|$$

$$+ \left| \int_{t^v(x_1,t)}^{t^v(x_2,t)} g(v(\tau, \xi^v(\tau(x_2,t), x_2, t)) \, d\tau \right|$$

$$\leq L_g \int_0^t |v(\tau, \xi^v(\tau, x_1, t)) - v(\tau, \xi^v(\tau, x_2, t))| \, d\tau$$

$$+ |t^v(x_1, t) - t^v(x_2, t)| \, \theta a^2$$

$$\leq L_g \int_0^t |v(\tau, \xi^v(\tau, x_1, t)) - v(\tau, \xi^v(\tau, x_2, t))| \, d\tau$$

$$+ \theta a^2 \frac{1}{a - \kappa} \exp(L_v T) |x_1 - x_2|$$

where we have used (6.61) from Lemma 6.5. Inequality (6.60) implies

$$f_2(t, x) \leq L_g \int_0^t |v(\tau, \xi^v(\tau, x_1, t)) - v(\tau, \xi^v(\tau, x_2, t))| \, d\tau$$

$$+ \theta a^2 \frac{1}{a - \kappa} \exp(L_v T) |x_1 - x_2|$$

$$\leq L_g T L_v \exp(L_v T) |x_1 - x_2|$$

$$+ \theta a^2 \frac{1}{a - \kappa} \exp(L_v T) |x_1 - x_2|$$

$$= [L_g T L_v + \frac{\theta a^2}{a - \kappa}] \exp(L_v T) |x_1 - x_2|.$$

Hence

$$L_2 = \left[L_g T L_v + \frac{\theta a^2}{a - \kappa} \right] \exp(L_v T)$$

is a Lipschitz constant of f_2.
To get a Lipschitz constant for f_1 we distinguish three cases.
Without loss of generality we assume that

$$t^v(x_1, t) \leq t^v(x_2, t).$$

Case 1.: If $t^v(x_1, t) > 0$, we have $\xi^v(t^v(x_1, t), x_1, t) = 0$ and $\xi^v(t^v(x_2, t), x_2, t) = 0$.
This implies

$$f_1(t, x) \leq |u(t^v(x_1, t)) - u(t^v(x_2, t))|$$

$$\leq \tilde{L}_R |t^v(x_1, t) - t^v(x_2, t)|$$

$$\leq \tilde{L}_R \frac{1}{a - \kappa} \exp(L_v T) |x_1 - x_2|.$$

Case 2.: If $t^v(x_2, t) = 0$, we have $t^v(x_1, t) = 0$ and $\xi^v(t^v(x_1, t), x_1, t) > 0$, $\xi^v(t^v(x_2, t), x_2, t) > 0$. This implies

$$f_1(t, x) \leq |v_0(\xi^v(0, x_1, t)) - v_0(\xi^v(0, x_2, t))|$$
$$\leq \tilde{L}_R |\xi^v(0, x_1, t) - \xi^v(0, x_2, t)|$$
$$\leq \tilde{L}_R \exp(L_v T) |x_1 - x_2|.$$

Case 3.: If $t^v(x_2, t) > 0$ and $t^v(x_1, t) = 0$, we have

$$f_1(t, x) = |v_0(\xi^v(0, x_1, t)) - u(t^v(x_2, t))|$$
$$\leq \tilde{L}_R |a\, t^v(x_2, t) + \xi^v(0, x_1, t)|$$
$$\leq \tilde{L}_R a\, |t^v(x_2, t) - 0| + \tilde{L}_R |\xi^v(0, x_1, t) - 0|$$
$$\leq \tilde{L}_R a\, |t^v(x_2, t) - t^v(x_1, t)| + \tilde{L}_R |\xi^v(0, x_1, t) - \xi^v(0, \tilde{x}, t)|$$
$$\leq \tilde{L}_R a\, \frac{1}{a - \kappa} \exp(L_v T) |x_1 - x_2|$$
$$+ \tilde{L}_R \exp(L_v T) |x_2 - x_2|.$$

Here $\tilde{x}$ is the point between x_1 and x_2 with $\xi_v(0, \tilde{x}, t) = 0$.

Case 1–Case 3 yield the Lipschitz constant L_1 for f_1 that is given by the equation

$$L_1 = \tilde{L}_R \exp(L_v T) \left[1 + \frac{a}{a - \kappa} \right].$$

Thus we get the Lipschitz constant L_Φ for $\Phi(v)$ that is given by

$$L_\Phi = L_1 + L_2$$
$$= \exp(L_v T) \left[L_g\, T\, L_v + \frac{\theta a^2}{a - \kappa} + \tilde{L}_R \left(1 + \frac{a}{a - \kappa} \right) \right]$$
$$\leq L_M$$

where the last inequality follows from our assumptions in Theorem 6.4. Therefore the Lipschitz constants are uniformly bounded by L_M during the fixed point iteration.

In Step 1 we have shown that the solution remains subcritical during the iteration. Hence for all $v \in M$ we have $\Phi(v) \in M$, that is

$$\Phi(M) \subset M.$$

Step 3: Contractivity Now we show that Φ is a contraction. For all $v, w \in M$ we have the inequality

$$|\Phi(v) - \Phi(w)| \leq A + I$$

with

$$A = |v(t^v(x,t), \xi^v(t^v(x,t), x, t)) - w(t^w(x,t), \xi^w(t^w(x,t), x, t))|,$$

$$I = \left| \int_{t^v(x,t)}^{t} g(v(\tau, \xi^v(\tau(x,t)))) \, d\tau - \int_{t^w(x,t)}^{t} g(w(\tau, \xi^w(\tau(x,t)))) \, d\tau \right|$$

$$\leq \left| \int_{t^v(x,t)}^{t} g(v(\tau, \xi^v(\tau(x,t)))) - g(w(\tau, \xi^w(\tau(x,t)))) \, d\tau \right|$$

$$+ \left| \int_{t^w(x,t)}^{t^v(x,t)} g(w(\tau, \xi^w(\tau(x,t)))) \, d\tau \right|$$

$$\leq L_g \int_0^t |v(\tau, \xi^v(\tau, x, t)) - w(\tau, \xi^w(\tau, x, t))| \, d\tau$$

$$+ |t^w(x,t) - t^v(x,t)| \, \theta a^2$$

$$\leq L_g \int_0^t |v(\tau, \xi^v(\tau, x, t)) - w(\tau, \xi^w(\tau, x, t))| \, d\tau$$

$$+ \theta a^2 \frac{1}{a - \kappa} T \exp(L_v T) \|v - w\|_{C([0,T] \times [0,\infty))}.$$

We have

$$I \leq L_g \int_0^t |v(\tau, \xi^v(\tau, x, t)) - v(\tau, \xi^w(\tau, x, t))| \, d\tau$$

$$+ L_g \int_0^t |v(\tau, \xi^w(\tau, x, t)) - w(\tau, \xi^w(\tau, x, t))| \, d\tau$$

$$+ \frac{\theta a^2}{a - \kappa} T \exp(L_v T) \|v - w\|_{C([0,T] \times [0,\infty))}$$

$$\leq L_g T L_v T \exp(L_v T) \|v - w\|_{C([0,T] \times [0,\infty))}$$

$$+ L_g T \|v - w\|_{C([0,T] \times [0,\infty))}$$

$$+ \frac{\theta a^2}{a - \kappa} T \exp(L_v T) \|v - w\|_{C([0,T] \times [0,\infty))}$$

$$= L_g T [1 + L_v T \exp(L_v T)] \|v - w\|_{C([0,T] \times [0,\infty))}$$

$$+ \frac{\theta a^2}{a - \kappa} T \exp(L_v T) \|v - w\|_{C([0,T] \times [0,\infty))}$$

$$= T \left[L_g (1 + L_v T \exp(L_v T)) + \frac{\theta a^2}{a - \kappa} \exp(L_v T) \right] \|v - w\|_{C([0,T] \times [0,\infty))}.$$

Now we consider again three cases. Without loss of generality we assume that

$$t^v(x,t) \le t^w(x,t).$$

Case 1.: If $t^w(x,t) > 0$, we have $\xi^v(t^v(x,t), x, t) = 0$ and $\xi^w(t^w(x,t), x, t) = 0$. Hence we get

$$
\begin{aligned}
A &\le |u(t^v(x,t)) - u(t^w(x,t))| \\
&\le \tilde{L}_R |t^v(x,t) - t^w(x,t)| \\
&\le \tilde{L}_R \frac{1}{a-\kappa} T \exp(L_v T) \|v - w\|_{C([0,T]\times[0,\infty))}.
\end{aligned}
$$

Case 2.: If $t^w(x,t) = 0$, we have $t^v(x,t) = 0$ and $\xi^v(t^v(x,t), x, t) > 0$, $\xi^w(t^w(x,t), x, t) > 0$. Thus we get

$$
\begin{aligned}
A &\le |v_0(\xi^v(0, x, t)) - v_0(\xi^w(0, x, t))| \\
&\le \tilde{L}_R |\xi^v(0, x, t) - \xi^w(0, x, t)| \\
&\le \tilde{L}_R T \exp(L_v T) \|v - w\|_{C([0,T]\times[0,\infty))}.
\end{aligned}
$$

Case 3.: If $t^w(x,t) > 0$ and $t^v(x,t) = 0$, we have

$$
\begin{aligned}
A &= |v_0(\xi^v(0, x, t)) - u(t^w(x,t))| \\
&\le \tilde{L}_R |a\, t^w(x,t) + \xi^v(0, x, t)| \\
&\le \tilde{L}_R a\, |t^w(x,t) - 0| + \tilde{L}_R |\xi^v(0, x, t) - 0| \\
&\le \tilde{L}_R a\, |t^w(x,t) - t^v(x,t)| + \tilde{L}_R |\xi^v(0, x, t) - \xi^w(0, x, t)| \\
&\le \tilde{L}_R a\, \frac{1}{a-\kappa} T \exp(L_v T) \|v - w\|_{C([0,T]\times[0,\infty))} \\
&\quad + \tilde{L}_R T \exp(L_v T) \|v - w\|_{C([0,T]\times[0,\infty))}.
\end{aligned}
$$

Here $\xi^w(0, x, t) < 0$ is defined by the extension of the characteristic curves for $(s, x, t) \in [0,T] \times (-\infty, \infty) \times [0,T]$. For this purpose w is extended on $(-\infty, 0) \times [0,T]$ by $w(x,t) = w(0,t)$ $(x < 0)$. Note that this extension is Lipschitz continuous.

From Case 1–Case 3 we get

$$
\begin{aligned}
&\|\Phi(v) - \Phi(w)\|_{C([0,T]\times[0,\infty))} \\
&\le L_{kontr} \|v - w\|_{C([0,T]\times[0,\infty))}
\end{aligned}
$$

with the Lipschitz constant

$$L_{kontr} = T\left[L_g \left(1 + L_v \, T \exp(L_v \, T)\right) + \frac{\theta a^2}{a - \kappa} \exp(L_v \, T)\right]$$

$$+ \tilde{L}_R \, \frac{\max\{a, \, 1\}}{a - \kappa} \, T \exp(L_v T) + \tilde{L}_R \, T \exp(L_v T).$$

Thus we have shown that if $T\left[\tilde{L}_R + L_g + \theta\right]$ is sufficiently small the map Φ is a contraction. Now Banach's fixed point theorem implies the existence of a unique fixed point of Φ in M, which is the solution of the quasilinear initial boundary value problem (*QARWP*). Thus Theorem 6.4 is proved. $\square$

Chapter 7
Distributions

For the analysis of partial differential equations often derivatives in the sense of distributions are needed, since classical solutions do not exist. Therefore we present a very short introduction to the theory of distributions that has essentially been influenced by LAURENT SCHWARTZ (see [50]). First we define the set of test functions. Let the dimension $n \in \{1, 2, 3, \ldots\}$ be given.

Definition 7.1. Let $\Omega \subset \mathbb{R}^n$ be open. We define

$$\mathcal{D}(\Omega)$$

as the set of test function $\varphi : \Omega \to \mathbb{R}$ with the following properties:

i) φ has compact support

$$\operatorname{supp} \varphi = \overline{\{x \in \Omega : \varphi(x) \neq 0\}}$$

that is contained in the set Ω (the notation $\overline{\{\cdot\}}$ denotes the closure of a set);
ii) φ is infinitely often differentiable.

Example 7.1. Let $n = 1$. Then

$$\varphi(x) = \begin{cases} 0 & \text{if } |x| \geq 1, \\ \exp(-\frac{1}{1-x^2}) & \text{if } x < 1 \end{cases}$$

defines a test function $\varphi \in \mathcal{D}(\Omega)$ with $\Omega = \mathbb{R}$ or $\Omega = (a, b)$ with $[-1, 1] \subset (a, b)$. We have $\operatorname{supp} \varphi = [-1, 1]$. Since $\varphi^{(n)}(1) = 0 = \varphi^{(n)}(-1)$ for all $n \in \{0, 1, 2, \ldots\}$, the function φ is infinitely often differentiable. Around the points $x = 1$ and $x = -1$ the function φ *cannot* be represented as a Taylor series.

© The Author(s) 2015
M. Gugat, *Optimal Boundary Control and Boundary Stabilization of Hyperbolic Systems*, SpringerBriefs in Electrical and Computer Engineering,
DOI 10.1007/978-3-319-18890-4_7

Analogously for $\Omega = \mathbb{R}^n$ we consider

$$\varphi(x) = \begin{cases} 0 & \text{if } \|x\| \geq 1, \\ \exp(-\frac{1}{1-\|x\|^2}) & \text{if } \|x\| < 1. \end{cases}$$

Exercise 7.1. Show that $\mathcal{D}(\Omega)$ is a vector space.

Exercise 7.2. Show that $\mathcal{D}(\Omega)$ is even an algebra.

Exercise 7.3. Prove the *approximation theorem*:
  Let $f : \Omega \to \mathbb{R}$ be continuous with $K = \text{supp} f$ compact and $\text{supp} f \subset \Omega$.
  For all $\varepsilon > 0$ and $d > 0$ there exists $\varphi \in \mathcal{D}(\Omega)$ with

$$\sup_{x \in \Omega} |f(x) - \varphi(x)| \leq \varepsilon$$

and

$$\text{supp } \varphi \subset K_d = \bigcup_{k \in K} \{x : \|x - k\| \leq d\}.$$

Proof of the approximation theorem (Solution of Exercise 7.3). Consider the function

$$\theta_1(x) = \begin{cases} 0 \text{ if } \|x\| \geq 1, \\ \exp\left(-\frac{1}{1-\|x\|^2}\right) \text{ if } \|x\| < 1. \end{cases}$$

For a real number $a > 0$ define the kernel

$$\theta(x) = \frac{1}{k}\theta_1\left(\frac{1}{a}x\right).$$

Then the support of θ is the closed ball around zero with radius a. We choose k in such a way that

$$\int_{\mathbb{R}^n} \theta(X)\, dX = 1. \tag{7.1}$$

We define φ as the convolution

$$\varphi(x) = \int_{\mathbb{R}^n} f(x-z)\, \theta(z)\, dz = \int_{\mathbb{R}^n} \theta(x-z)\, f(z)\, dz.$$

Now we show that for a sufficiently small φ has the desired properties.
  If $x \notin K_a$, for all $z \in K$ we have $\|x - z\| \geq a$, and thus $\theta(x - z) = 0$. For all $x \notin K_a$ this implies the equation

$$\varphi(x) = \int_{z \notin K} \theta(x-z)\, f(z)\, dz = 0.$$

Hence we have

$$\text{supp } \varphi \subset K_a.$$

Since θ is infinitely often differentiable, also φ is infinitely often differentiable.
Due to the choice of k that yields (7.1) we have

$$|f(x) - \varphi(x)| \leq \int_{\mathbb{R}^n} |f(x) - f(x - z)| \; \theta(z) \, dz.$$

Since f is uniformly continuous on the compact set K, the following statement holds:
There is a number $a > 0$ such that for all $\|z\| \leq a$ and for all $x \in K$ we have

$$|f(x) - f(x - z)| \leq \varepsilon$$

and thus

$$|f(x) - \varphi(x)| \leq \int_{\|z\| \leq a} |f(x) - f(x - z)| \; \theta(z) \, dz$$

$$\leq \int_{\|z\| \leq a} \varepsilon \, \theta(z) \, dz = \varepsilon.$$

If we choose $a \leq d$, then φ has all desired properties and the approximation theorem
is proved. $\square$

Notation. For a multiindex

$$\alpha = (\alpha_1, \ldots, \alpha_n) \in \mathbb{N}^n$$

we define

$$|\alpha| = \alpha_1 + \ldots + \alpha_n.$$

For $\varphi \in \mathcal{D}(\Omega)$ we define

$$D^\alpha \varphi = \partial_{x_1}^{\alpha_1} \ldots \partial_{x_n}^{\alpha_n} \varphi.$$

Definition 7.2. A distribution u on Ω is a linear map

$$u : \mathcal{D}(\Omega) \to \mathbb{R}$$

such that we have $\lim_{k \to \infty} \langle u, \varphi_k \rangle = 0$ for every sequence $(\varphi_k)_k$ of test functions such
that conditions i) and ii) hold, that are defined as

i) There is a compact set $K \subset \Omega$ with supp $\varphi_k \subset K$ for all k;

ii) For all $\alpha \in \mathbb{N}^n$ the sequence

$$(D^\alpha \varphi_k)_k$$

converges uniformly to zero.

The set of distributions is a linear space that is denoted as $\mathcal{D}'(\Omega)$.

Remark 7.1. Condition ii) in Definition 7.2 requires the uniform convergence for all derivatives of the φ_k. However, the uniform convergence is only required for each sequence

$$(D^\alpha \varphi_k)_k$$

separately and not simultaneously for all D^α.

Example 7.2. Let

$$f : \Omega \to \mathbb{R}$$

be locally integrable. Then by the equation

$$\langle T_f, \varphi \rangle = \int_\Omega f(x)\, \varphi(x)\, dx$$

f defines a distribution.

Proof. For a sequence $(\varphi_k)_k$, with $\varphi_k \in \mathcal{D}(\Omega)$, supp $\varphi_k \subset K$ and $\varphi_k \to 0$ uniformly for $k \to \infty$ we have

$$\begin{aligned}
|\langle T_f, \varphi_k \rangle| &= |\int_\Omega f(x)\, \varphi_k(x)\, dx| \\
&\leq \int_K |f(x)|\, |\varphi_k(x)|\, dx \\
&\leq \int_K |f(x)|\, dx\, \sup_{x \in K} |\varphi_k(x)| \\
&\to 0 \text{ for } k \to \infty.
\end{aligned}$$

Lemma 7.1. *Two functions f, g that are locally integrable satisfy the equation*

$$T_f = T_g$$

if and only if

$$f(x) = g(x)$$

almost everywhere.

Exercise 7.4. Prove Lemma 7.1!

7.1 Distributional derivatives

Now we want to define the partial derivative

$$\frac{\partial T}{\partial x_1}$$

with respect to x_1 for a distribution T in such a way that for any continuously differentiable function f we have

$$\frac{\partial}{\partial x_1} T_f = T_{\frac{\partial f}{\partial x_1}}.$$

First we consider an infinitely often differentiable function f. Then for any $\varphi \in \mathcal{D}(\Omega)$ we have

$$\langle \frac{\partial f}{\partial x_1}, \varphi \rangle = \int_{\Omega} \frac{\partial f}{\partial x_1} \varphi \, dX.$$

We extend φ by zero function values to a function that is defined on the whole $\mathbb{R}^n$. Since φ has compact support, Fubini's Theorem implies

$$\langle \frac{\partial f}{\partial x_1}, \varphi \rangle = \int_{x_2, x_3, \dots, x_n} \int_{-\infty}^{\infty} \frac{\partial f}{\partial x_1} \varphi \, dx_1 \, dx_2 \dots dx_n$$

$$= \int_{x_2, x_3, \dots, x_n} \left(-\int_{-\infty}^{\infty} f \frac{\partial \varphi}{\partial x_1} \, dx_1 \right) dx_2 \dots dx_n$$

$$= -\int_{\Omega} f \frac{\partial \varphi}{\partial x_1} \, dX$$

$$= -\langle f, \frac{\partial \varphi}{\partial x_1} \rangle.$$

Thus we have the equation

$$\langle \frac{\partial f}{\partial x_1}, \varphi \rangle = -\langle f, \frac{\partial \varphi}{\partial x_1} \rangle.$$

Therefore for a given distribution T we define the partial derivative

$$\frac{\partial T}{\partial x_1}$$

by the equation

$$\langle \frac{\partial T}{\partial x_1}, \varphi \rangle = -\langle T, \frac{\partial \varphi}{\partial x_1} \rangle$$

for all test functions φ.

Exercise 7.5. Show that $\frac{\partial T}{\partial x_1}$ is a distribution.

Now we consider a second order partial derivative

$$\frac{\partial^2 T}{\partial x_i \, \partial x_j}.$$

We have

$$\langle \frac{\partial^2 T}{\partial x_i \, \partial x_j}, \varphi \rangle = -\langle \frac{\partial T}{\partial x_j}, \frac{\partial \varphi}{\partial x_i} \rangle$$

$$= \langle T, \frac{\partial^2 \varphi}{\partial x_j \, \partial x_i} \rangle$$

$$= \langle T, \frac{\partial^2 \varphi}{\partial x_i \, \partial x_j} \rangle$$

$$= \langle \frac{\partial^2 T}{\partial x_j \, \partial x_i}, \varphi \rangle.$$

Thus we get the equation

$$\frac{\partial^2 T}{\partial x_i \, \partial x_j} = \frac{\partial^2 T}{\partial x_j \, \partial x_i}.$$

In this way we obtain partial derivatives of arbitrary order for any locally integrable function.

Example 7.3. Consider the HEAVISIDE-function

$$Y(x) = \begin{cases} 0, & \textit{if } x < 0 \\ 1, & \textit{if } x \geq 0. \end{cases}$$

For the derivative we have

$$\langle Y', \varphi \rangle = -\langle Y, \varphi' \rangle$$

$$= -\int_{-\infty}^{\infty} Y(x)\varphi'(x) \, dx$$

$$= -\int_{0}^{\infty} \varphi'(x) \, dx$$

$$= -(\varphi(x))|_{x=0}^{\infty}$$

$$= \varphi(0)$$

$$= \langle \delta, \varphi \rangle$$

with the famous DIRAC-distribution δ. Hence we have $Y' = \delta$.

For δ we get the derivative

$$\langle \delta', \varphi \rangle = -\langle \delta, \varphi' \rangle = -\varphi'(0)$$

and the higher order derivatives

$$\langle \delta^{(n)}, \varphi \rangle = (-1)^n \varphi^{(n)}(0).$$

Remark 7.2. In general two distributions S and T **cannot** be multiplied. However the product of an infinitely often differentiable function α and a distribution T can be defined by the equation

$$\langle \alpha T, \varphi \rangle = \langle T, \alpha\varphi \rangle.$$

With this definition, the following product rule holds:

Lemma 7.2.

$$\frac{\partial}{\partial x}(\alpha T) = \frac{\partial \alpha}{\partial x}T + \alpha\frac{\partial T}{\partial x}.$$

Exercise 7.6. Show that Lemma 7.2 holds.

Bibliography

1. Adams, R.A., Fournier, J.J.F.: Sobolev Spaces. Academic, Amsterdam (2003)
2. Avdonin, S.A., Ivanov, S.A.: Families of Exponentials. Cambridge University Press, Cambridge (1995)
3. Bennighof, J.K., Boucher, R.L.: Exact minimum-time control of a distributed system using a traveling wave formulation. J. Optim. Theory Appl. **73**, 149–167 (1992)
4. Bensoussan, A., Da Prato, G., Delfour, M.C., Mitter, S.K.: Representation and Control of Infinite Dimensional Systems. Birkhäuser, Basel (2007)
5. Birkhoff, G., Rota, G.-C.: On the completeness of Sturm-Liouville expansions. Am. Math. Mon. **67**, 835–841 (1960)
6. Bona, J., Winther, R.: The Korteweg-de Vries equation, posed in a quarter-plane. SIAM J. Math. Anal. **14**, 1056–1106 (1983)
7. Butkovski, A., Egorov, A.I., Lurie, K.A.: Optimal control of distributed systems (A survey of soviet publications). SIAM J. Control **6**, 437–476 (1968)
8. Butzer, P.-L., Berens, H.: Semi-Groups of Operators and Approximation. Springer, Berlin (1967)
9. Cerpa, E.: Control of a Korteweg–de Vries equation: a tutorial. Math. Control Relat. Fields **4**, 45–99 (2014)
10. Coron, J.-M.: Control and Nonlinearity. American Mathematical Society, Providence, RI (2007)
11. Coron, J.-M., Crepeau, E.: Exact boundary controllability of a nonlinear KdV equation with critical lengths. J. Eur. Math. Soc. **6**, 367–398 (2004)
12. Coron, J.-M., Lü, Q.: Local rapid stabilization for a Korteweg-de Vries equation with a Neumann boundary control on the right. J. Math. Pures Appl. **102**, 1080–1120 (2014)
13. Cox, S., Zuazua, E.L.: The rate at which energy decays in a string damped at one end. Indiana Univ. Math. J. **44**(2), 545–573 (1995)
14. Datko, R., You, Y.C.: Some second-order vibrating systems cannot tolerate small time delays in their damping. J. Optim. Theory Appl. **20**, 521–537 (1991)
15. Datko, R., Lagnese, J., Polis, M.P.: An example of the effect of time delays in boundary feedback stabilization of wave equations. SIAM J. Control Optim. **24**, 152–156 (1986)
16. Dick, M., Gugat, M., Herty, M., Leugering, G., Steffensen, S., Wang, K.: Stabilization of networked hyperbolic systems with boundary feedback. In: Leugering, G., et al. (eds.) Trends in PDE Constrained Optimization, pp. 487–504. Birkhäuser, Basel (2014)
17. Dieudonné, J.: Geschichte der Mathematik 1700–1900 – ein Abriss. Vieweg, Braunschweig/Wiesbaden (1985)

© The Author(s) 2015
M. Gugat, *Optimal Boundary Control and Boundary Stabilization of Hyperbolic Systems*, SpringerBriefs in Electrical and Computer Engineering,
DOI 10.1007/978-3-319-18890-4

18. Gugat, M.: Analytic solutions of L^∞ optimal control problems for the wave equation. J. Optim. Theory Appl. **114**, 397–421 (2002)

19. Gugat, M.: Optimal boundary control of a string to rest in finite time with continuous state. ZAMM **86**, 134–150 (2006)

20. Gugat, M.: Optimal energy control in finite time by varying the length of the string. SIAM J. Control Optim. **46**, 1705–1725 (2007)

21. Gugat, M.: Optimal boundary feedback stabilization of a string with moving boundary. IMA J. Math. Control Inf. **25**, 111–121 (2008)

22. Gugat, M.: Penalty techniques for state constrained optimal control problems with the wave equation. SIAM J. Control Optim. **48**, 3026–3051 (2009)

23. Gugat, M.: Boundary feedback stabilization by time delay for one-dimensional wave equations. IMA J. Math. Control Inf. **27**(2), 189–203 (2010)

24. Gugat, M.: Stabilizing a vibrating string by time delay. In: 15th International Conference on Methods and Models in Automation and Robotics, MMAR 2010, pp. 144–147, 23–26 Aug 2010. IEEE, Miedzyzdroje (2010)

25. Gugat, M.: Boundary feedback stabilization of the telegraph equation: decay rates for vanishing damping term. Syst. Control Lett. **66**, 72–84 (2014)

26. Gugat, M.: Norm-minimal Neumann boundary control of the wave equation. Arab. J. Math. **4**, 41–58 (2015) (Open Access)

27. Gugat, M., Herty, M.: Existence of classical solutions and feedback stabilization for the flow in gas networks. ESAIM: Control Optim. Calc. Var. **17**, 28–51 (2011)

28. Gugat, M., Herty, M.: The sensitivity of optimal states to time delay. PAMM **14**, 775–776 (2014)

29. Gugat, M., Leugering, G.: L^∞-norm minimal control of the wave equation: on the weakness of the bang–bang principle. ESAIM: Control Optim. Calc. Var. **14**, 254–283 (2008)

30. Gugat, M., Tucsnak, M.: An example for the switching delay feedback stabilization of an infinite dimensional system: the boundary stabilization of a string. Syst. Control Lett. **60**, 226–233 (2011)

31. Gugat, M., Leugering, G., Schittkowski, K., Schmidt, E.J.P.G.: Modelling, stabilization and control of flow in networks of open channels. In: Grötschel, M., et al. (eds.) Online Optimization of Large Scale Systems, pp. 251–270. Springer, Berlin (2001)

32. Gugat, M., Herty, M., Klar, A., Leugering, G.: Optimal control for traffic flow networks. J. Optim. Theory Appl. **126**, 589–616 (2005)

33. Gugat, M., Leugering, G., Sklyar, G.: L^p-optimal boundary control for the wave equation. SIAM J. Control Optim. **44**, 49–74 (2005)

34. Guo, B.Z., Xu, C.Z.: Boundary output feedback stabilization of a one-dimensional wave equation system with time delay, In: Proceedings of 17th IFAC World Congress, pp. 8755–8760 (2008)

35. Guo, B.Z., Yang, K.Y.: Danamic stabilization of an Euler- Bernoulli beam equation with time delay in boundary observation. Automatica **45**, 1468–1475 (2009)

36. Hörmander, L.: Lectures on Nonlinear Hyperbolic Differential Equations. Springer, Paris (1997)

37. Knobel, R.: An Introduction to the Mathematical Theory of Waves. American Mathematical Society/Institute for Advanced Study, Providence, RI (2000)

38. Komornik, V.: Rapid boundary stabilization of the wave equation. SIAM J. Control Optim. **29**, 197–208 (1991)

39. Krabs, W.: Optimal control of processes governed by partial differential equations part ii: vibrations. Z. Oper. Res. **26**, 63–86 (1982)

40. Krabs, W.: On Moment Theory and Controllability of One–Dimensional Vibrating Systems and Heating Processes. Lecture Notes in Control and Information Science, vol. 173. Springer, Heidelberg (1992)

41. Lasiecka, I., Triggiani, R.: Control Theory for Partial Differential Equations: Volume 2, Abstract Hyperbolic-like Systems over a Finite Time Horizon. Cambridge University Press, Cambridge (2011)

42. Leoni, G.: A First Course in Sobolev Spaces. American Mathematical Society, Providence, RI (2009)
43. Li, T.: Controllability and Observability for Quasilinear Hyperbolic Systems. AIMS, Springfield (2010)
44. Lions, J.L.: Exact controllability, stabilization and perturbations of distributed systems. SIAM Rev. **30**, 1–68 (1988)
45. Logemann, H., Rebarber, R., Weiss, G.: Conditions for robustness and nonrobustness of the stability of feedback systems with respect to small delays in the feedback loop. SIAM J. Control Optim. **34**, 572–600 (1996)
46. Pedersen, G.K. Analysis Now. Springer, New York (1989)
47. Rosier, L.: Exact boundary controllability for the Korteweg-de Vries equation on a bounded domain. ESAIM: Control Optim. Calc. Var. **2**, 33–55 (1997)
48. Rosier, L., Zhang, B.-Yu.: Control and stabilization of the Korteweg-de Vries equation: recent progresses. J. Syst. Sci. Complexity **22**, 647–682 (2009)
49. Russell, D.L.: Nonharmonic Fourier series in the control theory of distributed parameter systems. J. Math. Anal. Appl. **18**, 542–560 (1967)
50. Schwartz, L.: Méthodes mathématiques pour les sciences physiques. Hermann, Paris (1998)
51. Singh, T., Alli, H.: Exact time-optimal control of the wave equation. J. Guid. Control Dyn. **19**, 130–134 (1996)
52. Tröltzsch, F.: Optimale Steuerung partieller Differentialgleichungen. Vieweg Verlag, Wiesbaden (2005)
53. Toda, M.: Nonlinear Waves and Solitons. Kluwer, Dordrecht (1989)
54. Tucsnak, M., Weiss, G.: Observation and Control for Operator Semigroups. Birkhäuser Advanced Texts, Basel (2009)
55. Wang, J.-M., Guo, B.-Z., Krstic, M.: Wave equation stabilization by delays equal to even multiples of the wave propagation time. SIAM J. Control Optim. **49**, 517–554 (2011)
56. Whitham, G.B.: Linear and Nonlinear Waves. Wiley, New York (1976)
57. Work, D.B., Blandin, S., Tossavainen, O.-P., Piccoli, B., Bayen, A.M.: A traffic model for velocity data assimilation. Appl. Math. Res. Express **1**, 1–35 (2010)
58. Yosida, K.: Functional Analysis. Springer, Berlin/Germany (1965)
59. Zuazua, E.: Control and stabilization of waves on 1-d networks. In: Piccoli, B., Rascle, M. (eds.) Modelling and Optimisation of Flows on Networks, Cetraro, Italy 2009. Lecture Notes in Mathematics, vol. 2062, pp. 463–493 (2013)

Index

© The Author(s) 2015
M. Gugat, *Optimal Boundary Control and Boundary Stabilization of Hyperbolic Systems*, SpringerBriefs in Electrical and Computer Engineering,
DOI 10.1007/978-3-319-18890-4